NEW W...

BRIAN W. ALDISS
ERIC BROWN
PAT CADIGAN
GRAHAM CHARNOCK
WILLIAM GIBSON
PETER F. HAMILTON
NOEL K. HANNAN
GRAHAM JOYCE
GARRY KILWORTH
CHRISTINE MANBY
MICHAEL MOORCOCK
KIM NEWMAN
ANDREW STEPHENSON
HOWARD WALDROP
IAN WATSON

EDITED BY DAVID GARNETT

NEW WORLDS

Edited by David Garnett

Consulting Editor: Michael Moorcock

WHITE WOLF PUBLISHING
735 PARK NORTH BLVD.
SUITE 128
CLARKSTON, GA 30021
WWW.WHITE-WOLF.COM

CONTENTS

THE NEW WORLD'S NEW WORLDS

BY DAVID GARNETT

A true story: London, England, 1946. The year after the end of the Second World War. First publication of a new science fiction magazine. *New Worlds*. Edited by E.J. Carnell.

Since its original appearance, *New Worlds* has been through a number of incarnations. John Carnell edited the magazine for 18 years, and for a short time there was even an American reprint edition. (A very short time, five issues in 1960.) In 1964, at the age of 23, Michael Moorcock took over editorship of *New Worlds*—and shook up the whole multiverse of science fiction.

New Worlds was published as a monthly paperback for a few more years, before moving on to a larger format, and a gradually more erratic schedule. Ten volumes of *New Worlds* came out as paperback originals during the seventies, five of which also appeared in the USA as *New Worlds Quarterly*. There were a few more issues of the magazine. Then nothing. Until...

Meanwhile, I had edited two original anthologies, *Zenith* and *Zenith 2*. This was to have been an annual series, but the publisher was taken over by another and the series cancelled. Which was when Michael Moorcock asked if I would like to edit a new series of *New Worlds*. I would and I did, a four-book series being commissioned by the late Richard Evans of Gollancz. Richard knew the importance of new short fiction to the future of science fiction. He was an excellent editor, a good man, and he died far, far too young.

The series was published and met with an excellent critical response. Then nothing. Until...

White Wolf, who are reprinting all of Michael Moorcock's books in America, asked if I wanted to edit another series of *New Worlds*.

Which is what you are reading right now.

As it enters its second half-century, this is the first time *New Worlds* has seen initial publication in the USA.

From almost the very beginning, *New Worlds* has published stories by American authors. In a similar way, British writers might sell their stories to American magazines and their books might be published in America. (My first novel, for example, appeared in the USA before it found a British publisher.) This volume is being edited in Britain, and most of the contributors are British. Of the three American authors, Pat Cadigan recently moved to Britain, William Gibson lives in Canada—and Howard Waldrop's story is set in England. Michael Moorcock, however, now spends much of his time in the USA.

What British and American authors have in common, more or less, is the English language. It's an accident of history that Americans speak English. English is the language of Shakespeare—and of Hollywood. British films can be nominated for the Oscar (and even occasionally win). Because of the shared language, they are not considered "foreign."

America is the world's most powerful economy, and American media culture dominates the globe. English is the language of movies and television, music and advertising, comics and computer games, and so the world wants to speak English. English is the language of international trade, finance, commerce, diplomacy, and so the world has to speak English.

Despite the number of countries in which English is the mother tongue, it is not the world's most common first language. China is the most populous nation on Earth, and there are far more people who speak Chinese than English. But as a second language, the one people choose to learn, English has become the international lingua franca.

An apocryphal story: London, England, 1995. A young foreign visitor sees the 50th anniversary commemorations for the end of the Second World War, and he asks the tour guide who was fighting whom. "Britain, the Soviet Union and the United States were fighting against Germany and Japan," he is told. "Who won?" he asks, and the tour guide replies, "It's too early to tell."

Fifty years ago, *New Worlds* was not at all unusual. The majority of science fiction consisted of short stories published in genre magazines. There were very few SF novels, very few anthologies, the word "sci-fi" had not been invented; but there were a lot of magazines, most of them in the USA.

Now, *New Worlds* is very unusual. There are only a handful of American SF magazines still being published, while *Interzone* is the only one in Britain. But *New Worlds* has become an original anthology—which is even rarer than a science fiction magazine.

The science fiction short story itself is becoming a rarity. There are probably more SF novels published every year than short stories, although this stretches the word "novel" to extremes. A novel should be something new, original, unexpected; but there are very few of these any more.

It sometimes seems that the majority of new SF books are sequels (often to books by authors who have gone to the great remainder house in the sky) or the latest episode in an interminable series of novelisations.

Just as life imitates art, so literature imitates media. There are books based on films, on television, on adverts (yes, really), on comics, on computer games, on board games, on gaming cards. Thanks to its re-release, the biggest-grossing film of all time is *Star Wars*. (Although an American movie, it was made in Britain.) There was, of course, a book-of-the-film; and the two sequels were also "novelised."

More recently, new *Star Wars* novels have begun to appear with increasing frequency. This is an obvious attempt to follow the publishing success of *Star Trek*, and some authors write for both franchised series. There have been countless "original" *Star Trek* novels, producing sales figures which make the imprint the seventh largest publisher in the USA.

Several years ago, I spent a month touring the USA, and I checked out the television stations in a dozen states. The shows which were screened most frequently were *Star Trek* and *Cheers*. But there don't seem to have been any *Cheers* books. Is there no demand for titles such as "Cliff Loses a Letter" or "Woody's Vacation"? Why not an Early Years series, with books such as "Young Norm and Carla's First Date"?

As part of their marketing strategy, tie-in books are often issued to generate publicity for new films. The book-of-the-film is little more than a padded-out version of the script. (I know because I've "written" some myself.) But the phenomenon of a whole series of books based on a film or television show seems almost entirely restricted to science fiction.

Philip K. Dick's *Do Androids Dream of Electric Sheep*, for example, was filmed as *Blade Runner*. Now there are more *Blade Runner* novels. These are not written by Dick, however, who died in 1982. One of his short stories was filmed as *Total Recall*, and there was inevitably a novelisation of the screenplay. It must be time for a sequel. When will *Total Recall: the Forgotten Bits* be published...?

It's easy to understand why such books are published: because they sell. In the case of *Star Trek* and *Star Wars*, each volume sells hundreds of thousands of copies. It's less easy to understand why they do sell. A few of them, maybe. But a hundred *Star Trek* novels? Twenty, thirty, forty *Star Wars* novels?

And it's also understandable why "authors" write such things, because they get paid. Some of them get paid very well. Despite receiving a low percentage, they are producing a high-grossing product, which means they can often earn more than they would from writing a book of their own.

Star Trek novels have become assembly line fiction, mass-produced output from the fiction factory.

So, isn't everyone happy? Publishers make money, they pay their writers, and the readers get want they want: easy-to read-books.

In the short term, maybe everyone wins. But when it comes to the long march...

1946, the first year of *New Worlds*: Winston Churchill made his famous "iron curtain" speech about the division of Europe into communist and non-communist countries. (He was in the United States at the time.) The Second World War was over, but the Cold War had begun.

Fiction can be divided into two categories: novels and short stories.

There are still some books of short stories being published. These will be either single-author collections, which are usually reprints of stories which first appeared in the magazines, or else multi-author anthologies—which are also usually reprints of stories which first appeared elsewhere.

Reprint anthologies will always have a theme. Best of the Year, for example, in which the editor will make his or her choice of the "best" stories published the previous year. Or, say, Robot Serial Killers, in which the editor will choose from robot serial killer stories published in previous reprint anthologies. (The same stories tend to be recycled, on the basis that if they have appeared before they must be good enough to appear again. The editor thinks of a theme, checks out his computerised database, then up come the story titles. And there's another anthology.)

Nearly all original anthologies, books of new stories, also have a theme. A collection of lesbian cat vampire stories for example, all set in an alternate world where Abraham Lincoln married Queen Victoria. (If I'm exaggerating, it's not by much!) Theme anthologies are the only ones most publishers will produce. It's as if the reader wants to know what to expect before they even start a story: it must be something very like the previous one.

And perhaps this is what people really do want. What other reason is there for the success of Star Trek books, of trilogies, of sequels, of series? People like what they know, are reassured by the familiar. But that isn't what happens with New Worlds, where "new" means exactly what it says. There is no theme. All the stories are completely different. They have no connection with each other or with any which have appeared in previous volumes.

Apart from the fact that some of the authors are the same, the only link between this book and the last series of New Worlds is what you are reading now, my introduction. This is where I continue my "miserably rancorous" and/or "keenly perceptive" (according to Locus and Asimov's SF Magazine) editorials about science fiction.

Since New Worlds first appeared, many other magazines and anthologies have come—and most have gone, while all the other titles which preceded it have vanished.

New Worlds must be doing something right. By sticking to its policy of presenting the best new short stories, it has become the oldest continuing science fiction title in the world.

The first New Worlds was published after the end of the Second World War. Now, it's said, the Cold War is also over. Who won? It's too early to tell.

I'm writing this editorial on my personal computer, which was manufactured in the United States of America. The software for my word processor, however, is British.

Two weeks ago, I bought a new printer from an American company. On the back it says "Made in China."

China has the largest population in the world, and the Chinese People's Republic is still a communist dictatorship.

China also has the largest army in the world, but not everyone in the army marches and carries a gun. Factories are run on military lines, with recruits conscripted into industrial service. They live in barracks and produce consumer goods for export.

The United States gives China "most favored nation" trading status, and China has a huge and increasing balance of payments surplus with the USA.

And when they start producing *Star Trek* books, we'll really be in trouble.

David Garnett
Ferringshire, England
April 1997

THE EMPEROR'S NEW REALITY

BY PAT CADIGAN

for Storm Constantine

Once upon a time in a far-off land, there was an emperor who had unlimited free access to as many realities as he could ever want. This was because, being the emperor, he had free accounts with all the reality providers but, even more important, he also had the best hardware available, and the best software to run it, mostly things so far advanced that they wouldn't be available to regular people for at least a year. And this wasn't beta-test stuff, this was finished product that worked right, first time, every time. Nobody wanted to cheese off the emperor, who was not only the best advertisement a company could have for its goods, but a bottomless well of new notions and good ideas. For, whenever the emperor fancied having a new experience in a new way, he would phone up a hardware or software company and ask the R&D department to produce it for him.

The emperor had quite an imagination and was directly responsible for many different kinds of realities enjoyed by vast multitudes of people the world over. If these realities had been old-time theme parks, they would have been impossibly expensive to build or maintain, impossibly dangerous, and inaccessible to all but a very few. Fortunately, Artificial Reality™ had leveled *that* playing field all the way around; most people could afford the necessary hardware and software to enjoy at least a nice selection of realities on cable through affordable pricing plans, though no one had either the quantity or quality that the emperor had.

Now, while the people learned to make do even as they aspired to more and better, the emperor was not limited in what he called his 'quest for excellence,' and he tried to upgrade as much as possible every six months, sooner if he could manage it. As well, he studied aesthetics and developed an aesthetics of reality-building that became required reading at every institute of higher learning with a philosophy and/or design department, and caused several stubborn modernists to concede and embrace postmodernism. He immersed himself in the writings and work of Stanislavski, Strasbourg, and Olivier, and published a handbook on creating an on-line persona that was bundled free with every new hardware system or upgrade module sold during the following year. In between, he averted three wars, settled half-a-dozen strikes, and presided over the opening of a new mall in the capital city.

And still he made himself available on-line, to all of his subjects, in whatever reality he chose to manifest in. He did not always manifest as an emperor per se—sometimes he was a small boy, or a beautiful woman, or a racehorse, or some sort of fabulous monster that a girl named Alice might have encountered if she'd had a longer-lasting account with the Looking Glass world. But no matter what form he was in, there was always a special, forgery-proof tag that identified him as the emperor. (No one knows how R&D made it forgery-proof, not even the emperor; you'll just have to take my word for it.) The emperor explained that the tag wasn't there because he wanted everyone to bow down to him but because past experience had shown him that people who later found out they had been interacting with the emperor without knowing it complained of feeling spied on or tricked; apparently, suspicion of the government was one of those things that might as well have been genetic. So the emperor decided that honesty was the best policy.

After a time, the concepts of 'excellence' and 'honesty' became entwined in the emperor's mind. He believed that honesty must be at the core of all things excellent and so he called on hardware and software companies everywhere to strive for more and more excellence—i.e., honesty—in the realities they offered.

"The three most important things in the reality business," the emperor would say, "are authenticity, authenticity, authenticity. One should be able, if one wishes, to forget if one's activity is engaged in by way of hardware and software."

Eventually, this began to get on everyone's nerves. The emperor's subjects grumbled that it was all very well for someone with limitless resources to go on and on about authenticity and how *real* everything ought to be, but they had to be satisfied with what they could afford, and ration their time with it as well. The R&D departments at the hardware and software companies felt that they had pushed the realism stuff about as far as it would go. It was, after all, *Artificial Reality*™. Anything with a ™ after it was artificial; it wasn't *supposed* to be real, it was supposed to be an experience impossible to achieve in the real world. So how could anyone judge the authenticity of such a thing?

Realism, the emperor said earnestly. You have to *believe* it's real. If you can believe it's real, then it's *authentic*.

And so the people's senses were bombarded with authentic sensation the likes of which most of them had never seen, heard, felt, smelt, or tasted before in their lives. Everything in Artificial Reality™ was more real than real. Headmounted monitors achieved such resolution that the previously flawless skin of popular sex gods and goddesses acquired pores and even the oddly-placed hair or two. A kindergarten class on a virtual nature walk wandered into a virtual patch of poison ivy and came down with psychosomatic rashes that resisted all treatment save real calamine lotion. A woman became convinced that her own real life was the Artificial Reality™ she was visiting and vice versa, and the fact that the two were, for reasons known only to her, virtually identical didn't help.

But most people simply complained of feeling hammered on by the sheer intensity of everything. It was like asking for a snack and getting an eight-course dinner for nine, someone said, and someone else said, it was like needing a hug and getting ten hours of screaming sex. Yet another opined it was like wanting a light shower for your garden and getting eleven days of driving rain with gale-force winds and still another pointed out that this very human urge to one-up each other's comparisons was probably somehow at the heart of the problem in the first place and they should stop now before this reality became as unbearable as all the others.

The people could have just quit fooling around with Artificial Reality™, got rid of their hotsuits and their headmounted monitors, canceled their cable subscriptions and gone back to watching other people pretend on TV screens, but that would have been just another unfair extreme. They liked the old Artificial Reality™, it was part of their lives. They were entitled to their entertainment. Besides which, mass abandonment of an entire medium, even if possible, would have thrown a lot of people out of work and into poverty, a fate they didn't deserve just because the emperor had this authenticity fixation.

And they could have overthrown the emperor, except that no one could agree on how he should have been replaced. A bad choice in that area would ruin a lot more than just a lot of realities on cable.

Finally, a hardware company began to sell, very quietly, hotsuits that could be cooled down—i.e., the sensations could be dulled rather than intensified. That was a pretty good solution. It would have worked, except that the CEO of the company was greedy and made the 'suits proprietary, which is to say, they worked only with his company's hardware and were incompatible with anyone else's headmounted monitors or cable interfaces. When other hardware companies tried to license the specs for the 'suit, he refused and hung out at his desk, not returning phone calls and waiting for everyone to go replace their old systems with his so he could get richer. But everyone *didn't* go replace their old systems with his: Some bootlegged the specs, made 'suits, and sold them illegally. Others jerry-rigged something with variable results, and everyone agreed that when an emperor went bad in some way, there was always someone even worse.

Who knows how things would have deteriorated, except that a pair of philosophy students happened to reach a threshold. They were postmodernists, of course, and they had lately been engaged in a series of arguments with some other students concerning the role of hyper-realism in the postmodernist construct. The only issue anyone could agree on was that hyper-realism in Artificial Reality™ had to go.

"What's the point of Artificial Reality™ if it's *real?*" said one of the students, a handsome young woman named Sadie. She and the other student, Nick, were having a cola in the middle of a retro party called a rave, which was more peaceful than the Artificial Reality™ they had both just come out of. Things had become so intense in Artificial Reality™ these days that when people disengaged, they had to decompress by attending riots or wild parties, or risk a case of psychic bends. "The idea is to turn one's back on reality altogether."

"Annihilation?" Nick said, nodding significantly.

"I don't know," said Sadie, "I can hardly think for all the reality I've just been bombarded with. It doesn't matter. All reality everywhere will continue to be unbearable until we get the emperor off this hyper-reality jag."

"Annihilation?" Nick said again, sounding hopeful this time.

"Just the opposite, actually." Sadie's smile was even more wicked than usual. "If the emperor wants absolutely authentic reality, then someone should give it to him."

A week later, the emperor received a pair of entrepreneurs, new to the business of Artificial Reality™ and eager to show him something brand new in the way of hardware and software.

"Exalted One, what we have to present to you today is not just a matter of improved hardware and software," Sadie said in her most sincere voice. "It is positively revolutionary. My colleague and I have been working for years to achieve the excellence, the *authenticity* that you in your wisdom have called for. We believe that we have accomplished what no other R&D department in any other company has been able to do."

"And that is?" the emperor said after a long moment of silence.

"An experience so real as to be completely indistinguishable from a real experience." Sadie stood up a little bit straighter. "We call it *completely transparent reality.*"

Nick looked at her quizzically. This was the first he'd heard of *completely transparent reality.* That was because the phrase had only just popped into Sadie's head moments before. Sadie smiled reassurance at him. "My colleague and I will bring in the trunk now," she said, and turned to the emperor, who was on the edge of his throne. "With your permission, of course."

"And hurry," said the emperor. "*Completely transparent reality. Sounds delectable.*"

Sadie and Nick withdrew from the emperor's receiving room and returned a few seconds later carrying a shiny black trunk between them. It was actually a footlocker that they had borrowed from one of the ex-modernists, who had recently developed a sense of humor. As they set it down before the throne, the emperor jumped up eagerly and ran to open it.

"By your leave, Excellency," Sadie said, covering the latches with her own hands. "Please understand that this is unlike anything you have ever known before. It is not something that you, even with all your experience of more realities than most of us can even conceive of, should handle unaided. My colleague and I *must* assist you, for everything must be assembled in a certain order and connected in a certain way. There are so many complicated calculations and calibrations that must be made, and then double- and triple-checked in base two, five, ten, and fourteen mathematics. No one without our intimate knowledge and comprehensive training should even try to operate such a system. The result would be disastrous, it would be—" Sadie paused, searching for a word scary enough.

"Annihilation," Nick offered.

"Exactly," Sadie said. "It would annihilate your brain."

The emperor's face puckered with distress. "Well, if it's not *safe...*"

"Oh, it's perfectly safe," Sadie said quickly, "as long as we assemble it for you and you do *exactly* as we instruct you, during the assembly and calculations and calibrations."

"This doesn't sound like a reality I'*d* call *completely transparent*," said the emperor uneasily.

"Oh, but it is," Sadie assured him, careful to keep her voice calm and even. "It's just that this is our first working model. Later, as we perfect the process, we should be able to streamline the components and set programs to automatically calculate and calibrate. But I think you'll see, Excellency, after you are completely wired up and ready to go, that it is definitely worth the trouble and effort. And that it *is*, without a doubt, *completely transparent reality.*"

"Well, all right," the emperor said. "What shall I do? Do I need to disrobe?"

Sadie and Nick looked at each other for a clue. "No," Sadie decided after a moment. "The, uh, hotsuit is so sensitive that your clothing will be completely transparent to it."

"Well, to *it*, maybe, but not to *me*. I get terribly overheated in a 'suit unless I'm a hundred percent naked—"

"Our revolutionary cooling system will make sure that not one drop of sweat will be necessary," Sadie said in a burst of inspiration. Behind the emperor, Nick rolled his eyes with relief. Sadie knew he must have been picturing their annihilation by a very naked and extremely angry emperor.

"What you must do, Excellency," Sadie continued, "is stand here—" she moved him so that he was standing behind the trunk with his back to it— "hold your arms slightly away from your body, and close your eyes."

"Like this?" said the emperor, unconsciously mimicking the stance of an emperor penguin.

"Perfect," Sadie said as Nick opened the trunk. "No doubt you'll be helping us fit this kind of 'suit to other people in the future, I think you have a knack for it. Now, I must caution you, Excellency, that whatever you do, you *must not open your eyes*. Until we tell you it is safe, that is. The first thing we do is fit the headmounted monitor. There are ten times as many lasers to calculate and calibrate and it would be too easy to scratch the cornea if you happened to open your eyes at the wrong moment. Or worse, to burn some rods and cones."

"Well, all right," said the emperor, sounding uneasy again. "I promise I won't open my eyes."

"Very good, Excellency. We will take some measurements and begin."

And then Sadie and Nick stood back from the emperor and were careful to remain completely silent and motionless.

After a minute, the emperor looked troubled. His brow furrowed and his mouth began to twitch. He made a small noise in his throat and swallowed. Sadie saw his weight shift slightly but it was another minute before he said, "Excuse me, are you still there?"

"Yes, Excellency," Sadie said, moving smoothly and silently to speak directly into his right ear. "I am, even now, just finishing with your headmounted monitor fittings. You see—uh, you feel how lightweight it is."

"I'm wearing a helmet?" the emperor said, amazed, and started to reach up to his head. Nick lunged and caught his hands just in time.

"No, Excellency, you're wearing a headmounted monitor," Sadie said carefully. "But it is a headmounted monitor like no other in the world—any world—and I have not yet adjusted the thousands of laser settings inside, so you must not open your eyes at the risk of being blinded, and you must not attempt to touch the unit, as the placement must be absolutely precise."

"Oh. Sorry." The emperor obediently lowered his hands and returned to his penguin position. "I won't move unless you tell me, then."

"Thank you, Excellency," Sadie said, risking a sigh of relief. "That will make things so much easier. Your total cooperation will be rewarded beyond your wildest dreams."

Nick rolled his eyes again and made a lowering gesture with one hand, meaning that she should tone it down. *Annihilation*, he mouthed at her.

"Excuse me?" said the emperor, frowning a little, eyes still closed. "Did somebody say something?"

"No, no," Sadie said quickly, waving for Nick to step back. "I'm just calculating and calibrating the sound assembly and it must have tickled your eardrums in the process of self-testing. Don't worry, Excellency. It's all quite normal."

"All right. It's just—"

Sadie and Nick held their breath. "Yes?" Sadie said.

The emperor was twitching his mouth and wrinkling his nose. "Well, it's silly."

"Please, Excellency, if you have some concerns, you must tell us."

"My nose itches."

"Ah," Sadie said as Nick fanned himself and blew out a silent breath. "Please hold perfectly still while I adjust the equipment so that I can help you with that." She stood in front of him, counted to fifteen, and then scratched his nose for him. "Better now, Excellency?"

"Perfect. Continue."

She stepped away from him again, and she and Nick kept quiet and still, watching the emperor. After five minutes, she and Nick tiptoed to stand on opposite sides of him. "Excellency, you may feel a very slight but not unpleasant pressure at the back of your head." She waited another fifteen seconds. "There. That doesn't hurt, does it, Excellency?"

"Not a bit," the emperor said. "It was just as you said it would be— very slight, but not unpleasant."

Sadie blinked, glancing at Nick. "Do you still feel it?"

"Yes. Is that bad?"

"No, no, that's all right." She covered her mouth for a moment, afraid she might laugh. "The sensation will fade after a while. Just the—the interior of the headmount customizing itself to your head."

"Well, it certainly is comfortable. Not to mention light. Practically weightless."

"Practically," Sadie agreed. "Now, Excellency, in the next several minutes, you may hear a number of different sounds, and smell a variety of different, uh, aromas. You may even feel varying degrees of heat or cold, all over or just in spots. It's just the self-testing mechanism, making sure that the full range of experience is available. We may have to make some adjustments based on the readings. And now you must hold *perfectly* still and not move, not even to talk— you should even breathe as shallowly as you can—so that your hotsuit can be fitted."

"Not yet," said the emperor, sounding worried.

"Is something wrong?" Sadie asked him, hoping she didn't sound as panicky as she felt.

"I just need to yawn first before you do that."

"Please," Sadie said pleasantly, making a face, "take all the time you need, and let us know when you're ready for the hotsuit fitting."

The emperor lowered his arms to his sides, bent his elbows up, and yawned deeply. "Will I have to stand like this much longer? With my arms out, I mean. I'm getting some strain in my shoulders."

"We'll work as quickly as we can, Excellency," Sadie said, starting to feel a little guilty and annoyed at herself for it. After all, it wasn't as if a little shoulder-strain was going to kill him. And it was nothing compared to the strain everyone else had been enduring thanks to his authenticity crusade. "Ready?"

The emperor gave one last enthusiastic exhale. "Go to it."

She and Nick went to the trunk and took off their outer clothing. Sadie shook out the astronaut pressure suit and helped Nick into it. Then she picked up what looked like a rose blossom the size of a large pumpkin. In fact, it was a ski-mask slightly stiffened to allow rose-colored paper petals to be glued on. Nick made a face as she offered it to him.

Don't start, she mouthed at him.

He fanned himself again, pulling at the neck of the astronaut suit.

She looked pained. *You promised. You did.*

Nick looked down at the floor unhappily.

"Um, I'm not sure I hear anything," the emperor said something, "but I'm beginning to feel slightly warmer, and I'm sure I smell something."

Sadie gave Nick a dirty look and plucked one of several bottles of cologne out of the trunk. "Yes, Excellency? Does it smell like turpentine or fresh fish?" She sprayed a general mist of Crazy Gardenia over his left shoulder. The paper petals on the headpiece rustled as she moved.

"Well, actually, um, maybe, um, I don't know, neither. Well, more fish than turpentine, and maybe not all *that* fresh. But now I hear something!"

"Yes, Excellency?" Sadie said nervously, trying to make Nick take the headpiece. "Is it a waterfall or a sort of knocking?"

"It's a forest fire," the emperor said, sounding pleased. "Very distant, but I know that's what it is. I hear a forest fire."

"Amazing," Sadie said, trying to force the headpiece over Nick's head while he tried to hold her off. "Your hearing must be much more sensitive than we realized. This will take some extra calculating and calibrating—" she grunted, twisting away from Nick and shaking the headpiece at him angrily.

"Everything all right?" the emperor asked, concerned.

"Perfectly, yes, just great, Excellency. This is hard work, make no mistake, and sometimes *doing something the right way demands almost superhuman efforts under conditions of great difficulty.*" She held the headpiece out to Nick, who was sulking. "But that's how it is *if you want things to work right.*"

Nick snatched the headpiece from her and put it on, working it down over his head without any effort to keep his movements silent.

"That forest fire sounds a lot closer now," the emperor said. "I can all but feel the heat on my back." He paused. "Yes, there it is. I feel like I'm standing in front of an enormous fire, all right."

"Well, it should cool down in a minute, Excellency," said Sadie. "More like thirty seconds, actually. It's just the self-test, so everything is very brief." She waited, keeping an eye on him as she pulled her own costume out of the trunk. "There. Is it any cooler now?"

"Yes, much. Thank you." The emperor sniffed. "Ah. It's hailing. I'd know the sound anywhere."

The millions of tiny beads on the ball gown rattled and shook as Sadie climbed into the dress and got the off-the-shoulder neckline situated off her shoulders. She would have preferred to put Nick in the dress but it was too small for him and they had both been afraid that the emperor would not have trusted his perceptions. For all his techno-lust, he had a hard retro streak that could not be ignored.

"Do you smell bacon cooking yet?" Sadie asked him as she dabbed blue paint onto her face, watching herself in the mirror Nick was holding for her. *You look great,* she mouthed at him. *It's perfect.*

He flipped down the petal near his right eye to glare at her. She blew him a kiss. His costume had been put together very quickly on the spur of the moment after she had seen a painting of an astronaut with a rose in place of his head. The execution in life was nowhere near as romantic or haunting as the painting—they couldn't get the right kind of space suit, for one thing, and for another, the paper petals just didn't have quite the same texture. But as an image alone, it was spooky enough for their purposes.

The emperor sniffed again. "Is that fur?"

Sadie finished painting her face blue and started gluing dominoes to her collarbones. "You smell fur, Excellency?"

"Like an animal. A dog, or a—" he sneezed. "Oh, it's a cat. I'm so allergic to cats—oh, get down, you miserable creature." He sneezed again.

"They always go right for the one that's allergic to them, don't they?" Sadie said, inserting chrome contact lenses.

"I can feel it walking back and forth over my shoulders as if I'm a sofa or a back fence or something," said the emperor, sniffling. "Animals love me, *all* animals. Even animals I don't love love me. What's an emperor to do? Nice kitty, pretty puss. Now get *down*."

Sadie hesitated, looking up at Nick. She was starting to get worried. They weren't even done putting on his nonexistent hotsuit and he was having experiences as vivid and detailed as the old-style hotsuits had provided, before the intensity had been amped up to unendurable. Nick shrugged and motioned for her to finish up quickly. She nodded. They had gone too far now to turn back or call it off.

"That's a relief," said the emperor, sighing. "Oh, I'm sorry. I wasn't supposed to breathe too deeply, was I. Did I ruin anything?"

"No, Excellency, that's just fine. We're getting close to the end now."

"Can I open my eyes yet?"

"No, not yet. Still calculating and calibrating." She picked up a spray bottle of porcelain cleanser and squirted the air two feet in front of his face. "Tidal wave!"

The emperor flinched slightly. "Good God, that ocean air! I can feel the sand between my toes!"

Sadie frowned and took a close look at the label. Sure enough, it was *Sea Breeze Special*. She tossed it to Nick, who put it back in the trunk and closed the lid carefully.

"All right, Excellency, we're very close to the end now. You may feel a series of unrelated sensations of varying lengths and intensities. Don't be alarmed, it's just the final self-test sequences running. Although please let me know if you find any of it painful, of course." Sadie and Nick stood still again and watched the emperor.

"When did you put the hotsuit on me?" the emperor asked.

"While you were breathing shallowly," Sadie told him. "You were absolutely perfect, Excellency, you really were. We managed to get you an exact fit."

"Well, I can't wait to see this," the emperor said feelingly.

"You'll be amazed, Excellency. In fact, you may not believe your eyes."

Nick adjusted the petals in front of both eyes and gave Sadie a baleful stare.

"Oh, now, *that* I don't know about." The emperor chuckled. "I've seen a thing or two, you might say."

"All right, Excellency. Now see this."

"Now?" he asked. "I can open my eyes now?"

"*Carefully*," Sadie warned. "You may experience some very minor vertigo."

The emperor raised his eyelids as if he were lifting two window shades by force of mind alone. "Whoa," he said, putting out his arms and balancing himself with one foot behind him. "Yes, you're right about the vertigo. Brief but *so* intense." Then he looked around the room. "This is my receiving office." He caught sight of Nick and raised his eyebrows. Sadie moved to Nick's side. The emperor's gaze moved rapidly from Sadie to Nick and back several times. "Well. That's more like it. Very nice." He looked around the room again. "But this is just my receiving office."

"Amazing, isn't it," Sadie said. "When have you seen anything reproduced with such authenticity?"

The emperor looked at her, startled. "You're right. The authenticity is—it's—well, I'm at a loss for words." He turned to look at his throne. "Amazing. It's my throne down to the last detail. Even—" He went up the three steps to look at the left arm. "Yes, there it is. I put a tiny little nick just at the spot where my elbow sits, just because I wanted to leave my mark on it for the next emperor. I was very young when I did that, of course." He laughed at his younger self's folly. "Well. Amazing. It's just like being in my receiving room. I can feel the floor perfectly. I can smell the air—" he paused. "Has someone been spraying perfume or air freshener in here?"

"Just the olfactory array settling into its final configuration," Sadie told him, squeezing Nick's hand behind her back.

The emperor turned his head from side to side. "And this helmet! I can't even *feel* it! It's just like my own actual head—" he put his hands up to feel, clasped the sides of his face and jumped, looking down at himself. "Why, I can't feel the helmet or the hotsuit *at all!*"

"Is the realism too intense, Excellency?" Sadie asked, wide-eyed. "Is it too startling for you?"

The emperor's mouth opened and closed several times but nothing came out. He stumped down the steps from the throne, arms folded, his expression troubled. "Actually, I can't say that the realism is too *intense*. But it is awfully...well...*real*."

"That *does* take some getting used to," Sadie commiserated.

"It's not that it's *more* real than real," the emperor said, more to himself than to Sadie and Nick, "it's that it's *as* real as real. It's—it's—"

"Completely transparent reality," said Sadie.

"*On the money!*" said the emperor, clapping his hands, and then looking at them carefully. "Wow. I really felt that. Skin on skin. I felt it and heard it just the way I would have if I had actually done it for real. This is the lightest, least intrusive hotsuit I have ever had acquaintance with."

Sadie's eyes glittered. "Yes, Excellency. It is, indeed."

The emperor went over to Sadie and Nick and bowed to them. "I salute you. This is unprecedented, and without equal. I want to show this to as many people as possible right away. Tell my administrative assistant to come in here. We must call every form of media, every CEO of every hardware and software company, as many of their R&D people as possible, philosophy departments, cybernetic culture research units, designers—and get this on the Internet, right away—" He paused. "Should I take this off? No, no, I've barely begun the experience." He looked to Sadie. "You'll keep me from walking off a cliff, though, won't you."

"Of course," Sadie said.

The news traveled through the empire at the speed of light—the emperor had found the ultimate unobtrusive Artificial Reality™ equipment and would demonstrate it immediately at the grand assembly hall in the middle of the capital city. A rumor that this was a hoax akin to the Good Times virus followed promptly and then was squashed by confirmations from several media watch groups that the emperor was indeed on his way to the grand assembly hall; he had demanded the presence of every major techhead and postmodernist who could turn up without undue hardship—medium hardship didn't count—as well as anyone else who had any interest in Artificial Reality™; and it did have everything to do with a major technological breakthrough.

Everyone was used to the emperor broadcasting his enthusiasms to all and sundry, but this was unprecedented. Though he was as thoughtless as any other despot, benevolent or otherwise, the emperor was not the sort of person who demanded that everyone drop everything and get on over for a major announcement. Announcements went faster via the rumour-mill.

So those who were not obligated by way of their jobs came out of curiosity, and as a welcome relief from whatever Artificial Reality™ they had been suffering through. It could have been worse, they told each other. The emperor could have demanded that they all meet online, in some too-real-to-be-real scenario that would have them all taking extra doses of post-traumatic stress tranquilizers for months after.

Meanwhile, the emperor waited in the green room behind the stage in the largest auditorium, with Sadie and Nick, who had been unable to figure out any graceful way to absent themselves from the next bit. Neither of them had expected the emperor to declare a media sensation, although in retrospect, Sadie didn't understand how they could have been so blind. Authentic Artificial Reality™ had been the emperor's one and only passion for a long time. The two of them sat close together on a couch, clutching each other's hands, while the emperor wandered around the room touching the walls and the light fixtures, occasionally stopping in the lavatory to look at himself in the mirror and marvel at the quality of the reflection. "They *never* get mirrors right in these things," he murmured wonderingly. "But look at *this*. Just *look* at it. And still *completely transparent*."

Sadie shook her head sadly and looked up at Nick. "Oh, boy," she whispered. "What now?"

"Annihilation," Nick said with conviction.

"You're probably right." She sighed. "We're going to publish *and* perish."

The telephone trilled delightfully, as all telephones did. Converting telephone ringers from annoying to delightful had been an earlier passion of the emperor's, before he had discovered the joys of multiple realities. "Can I answer that in my 'suit and helmet?'" the emperor asked Sadie, pointing at the phone on the desk against the near wall.

"Absolutely," Sadie said. "You'll even feel the phone against your—" Nick elbowed her in the ribs and she shut up.

"Yes?" the emperor was saying. "Full? Really? In only two hours? That's marvelous. Yes, turn on the screens outside in the parking lots, that's a brilliant idea. We'll be right up." The emperor replaced the phone receiver and turned to Sadie and Nick, rubbing his hands together excitedly. "Well, it's showtime. The entire grand assembly hall is filled to capacity. They're going to broadcast on the outdoor screens in the parking lot, on every major network, and in every reality." He paused and looked at his hands, pressing them together, pulling them apart, rubbing the heels together, then the palms. "Amazing. Just amazing. It really feels like my hands!"

Sadie's smile was more like a grimace. "Well, we try not to write any checks we can't cash." Nick elbowed her again. "Oh, what does it matter?" she muttered.

"Not a bit, in a cashless society," said the emperor and chuckled. "All right, come on, it's upstairs to the grand assembly hall stage, so that everyone can take a look at this marvelous new development—" He clapped his hands and hustled them up off the couch and out of the room to the elevator down the hall.

Just as he was about to get on with them, he stopped, reached in, and pressed the button for the main floor. "You guys ride. I'll meet you up there."

"Excellency?" Sadie said, alarmed.

"I want to feel *stairs*," the emperor said and nipped out just as the doors closed.

Sadie and Nick fell into each other's arms.

"We're gonna die," Sadie said. "*Horribly.*"

Nick moved to take off the rose head but Sadie stopped him. "Leave it. There could be a miracle and we might escape. You wouldn't want your face broadcast everywhere." He made a disgusted noise but left the rose head on anyway.

Sadie had been hoping to slip out when the elevator stopped but assembly hall staff were waiting to escort them to the stage, where the emperor had two chairs on either side of his own public-appearance throne. Each was escorted to a seat where they sat facing what looked like an ocean of people. *We're going to get hated to death*, Sadie thought and looked at Nick. Nick stared straight ahead and she remembered how he had always maintained that he wanted to meet the Apocalypse head on and look it straight in the eye.

Abruptly, the emperor bounced onto the stage from the wings, looking flushed and out of breath, arms high in the air as if he had just won a marathon and was taking a victory lap. He actually did one jogging turn around the stage before coming to a stop in front of his throne.

"Ladies, gentlemen, and everyone in between," he panted, and the sound engineer took the stage mikes up several notches in volume, "behold the future!" Still holding his arms in the air, he turned slowly, displaying himself to everyone. He was still wearing his office receiving suit and it was starting to get a bit sweaty; Sadie caught a whiff of perspiration from where she was sitting. The emperor looked at her and winked. "Such realism!" He grabbed the front of his shirt. "I'm *sweating* like a *pig*." She spread her hands helplessly and looked past him to Nick who mouthed *Annihilation* at her before turning around to face front again.

"This *is* the revolution!" the emperor was saying. "There has *never*, in the history of reality, artificial or otherwise, been such sophisticated technology. What you see before you is your emperor arrayed in *completely transparent reality-ware!*"

He paused, waiting for some sort of reaction. There were a few whispers near the back but otherwise the great assembly hall was all but silent. The emperor looked down at himself and then at Sadie. She floundered for a few moments. "Maybe they just can't see it very well," she blurted finally.

"Damn, you're *right!*" said the emperor, and headed for the front of the stage.

Nick made a sarcastic face at her and gave her an equally sarcastic thumbs-up. She threw up her hands. If it was all annihilation anyway, what did it matter?

"All right now," the emperor said to the audience. "Watch *this.*" He jumped down from the edge of the stage onto the floor directly in front. Sadie closed her eyes, wincing at the sound of him hitting the carpet.

"Ouch!" announced the emperor and stood up. "Believe it or not, I really felt that one. I mean, I *really* felt it. Just like I can *really* feel *this*—" he slapped one hand on the stage "—and *this!*" He bounded back and forth in front of the first row as if he were doing a gazelle imitation. "I just walked up *stairs* and I felt *each and every one!* And do you know *why?*" He looked questioningly at the first few rows of people in front of him, who were now all wearing the same uneasy expression. "Come, come now," the emperor said, snapping his fingers and then pausing for a second to make a delighted noise at his hand. "Who knows why? Who wants to tell me why?" His head swung from side to side and then he pointed to the person seated directly in front of him. "Would *you* like to tell me why?"

The woman tried to disappear into her seat cushion and failed. "Um, would that be because you had just walked up stairs?"

"Ah, but I didn't *just* walk up stairs," the emperor said, raising one finger in that way, "I walked up stairs in *completely transparent reality-ware!*"

"Oh," said the woman and nodded to the people on either side of her. "Of course. That makes all the difference."

"You bet it does," said the emperor, parading up and down before the front row again. "This is an experience that is *as real as real*. Not *more* real. Not *un*-real. Not *surreal*. But *completely, utterly, apparently, transparently real.*"

"Uh-oh," said the front row. The words swept back through the entire audience, fanning out to where people were standing in the aisles, out the doors to the lobby, out of the lobby into the parking lot, into radio and TV transmissions, into online services and out over the Internet both in the encrypted and unprotected forms. Everyone within any sort of data transmission range experienced the same satori almost at the same moment: *The emperor has finally cracked.*

Now, if he'd been merely a governor or a president, he would have been declared incompetent and packed off to some asylum, and the lieutenant-governor or vice-president would have carried on. If he'd been a king or a prince, insanity would have already been accepted as a given and it would have been business as usual. But he was an emperor, which meant that he not only had all the power possible, he actually *used* it. He could close schools, hospitals, businesses, airports, railroads, TV stations, reality service providers; he could cut off food and energy supplies; he could call in napalm strikes or strafing runs. He never had in the past, but then, he'd never lost his mind before. There was no telling what he would do, and it was no good just telling the people around him who would obediently close schools, hospitals, etc., cut off food and energy supplies, and perform the napalm strikes and strafing runs not to obey the emperor because he was crazy—those people were all the product of vocational institutes for the education and training of underachievers and the otherwise inadequate. They were all programmed to do the emperor's bidding and *only* the emperor's bidding. They didn't know about insanity, they only knew they couldn't obey anyone *but* the emperor, and they could disobey anyone *except* the emperor.

And as far as emperors went, this one was much nicer than most, but a mad emperor was a dangerous emperor, no matter how nice he had been before he'd lost his marbles. A dangerous emperor had to be handled very carefully; until something could be figured out, a mad, dangerous emperor had to be *humoured*. Everyone turned to everyone else and whispered, *Just nod and smile at whatever he says so we can get out of here alive, okay?* Everyone, that was, except Sadie and Nick, who weren't sitting close enough to anyone to whisper, weren't getting any sort of transmissions, and were the only other people, besides the emperor himself, who didn't know what was going on. But they could hear the whispers, like the rustling of a thousand paper rose petals, or maybe the biggest forest fire there had ever been, and they were both positive the people in the audience were planning their deaths. Horribly.

So as the emperor ran up and down the aisles turning the occasional handspring, doing the odd dance step, and sniffing necks and wrists for traces of cologne or whatever, people remained very still, so as not to attract any attention to themselves, hoping that the emperor would soon work off his manic phase and drop into a stupor or something so he could be removed without muss or fuss.

But the man's energy seemed to have no bounds at all. Every time it seemed that he might be winding down, he would instead get rev up and run around the auditorium again, inviting people to touch his arms and head so that they could see how light and flexible the hotsuit was, how the helmet was positively weightless.

"Possibly *lighter* than air!" he crowed to one startled grande dame whose hand he was forcing through his salt-and-pepper curls. "Lighter than *hair!*" As he straightened up, laughing at his own joke, he caught sight of Sadie and Nick sitting frozen on the stage. In his enthusiasm for completely transparent reality-ware, he had all but forgotten them. "And *those* are the geniuses who developed completely transparent reality, and the completely transparent reality-ware that goes with it!" The audience turned as one to see whom he was pointing at. *So you're the ones who drove the emperor crazy*, the audience seemed to say silently. *You're dead meat.* Sadie was positive she heard someone actually mutter the words *dead meat* aloud.

"We did it for free!" she suddenly shouted, desperate. "We didn't make a penny on this! And we refuse to take *any* royalties! We work *pro bono!*"

"However, movie rights *are* negotiable," said the emperor and laughed. "How about that, everyone? Are they saints or are they saints? They are *donating* this marvelous technology, this revolution, this breakthrough, to humanity. Complete altruism! Surely we can all spare such benefactors a little for movie rights! Right? Right?" He looked around for support and pointed at an older man sitting half a dozen seats from the aisle. "Right?"

"They should get what they deserve," said the man. "Absolutely." He turned to give Sadie a murderous look.

"That's *right!*" said the emperor joyfully, skipping down the aisle. "I can hardly believe this, do you know that I can actually *smell* with this technology? I mean, *smell spontaneously!* No programming, no adjusting the settings, *it just happens!* Just like in real life!"

As he reached the front of the auditorium, an older woman sitting on the end seat in the front row stood up and moved to stand directly in front of him. "This *is* real life, you damned fool."

The entire audience, including everyone at home, gasped, and waited. The emperor stood staring at the woman in what looked like shock. Everyone waited for the shock to turn to rage. They seemed to wait forever. Then the emperor turned to Sadie and gave an incredulous laugh.

"*How* do you do it? *She looks and sounds exactly like my mother!*"

"For God's sake, Gerald, I *am* your mother!" the woman called after him as he skipped past her and hopped up onto the stage again. "Gerald, you turn around and *look at me when I'm talking to you—*"

But the emperor continued to walk back and forth across the stage, extolling the virtues of completely transparent reality, occasionally pointing at the woman who claimed to be his mother and praising her utter realism. Suddenly, Nick stood up, walked over to the emperor and hit him hard on the back of the neck. The emperor went down like a collapsing building.

The audience jumped to its feet and the emperor's mother said, "What the hell did you do *that* for? He's a damned fool but he didn't deserve—"

"When he wakes up," Nick said, "we will explain that the system malfunctioned, causing the sensation of being struck hard enough to lose consciousness. But the sensation was so real that he actually did lose consciousness. This means that completely transparent reality-ware is far too dangerous for public consumption and we have to go back to the drawing board with it, and it may take years and years of hard work, calculations, calibrations, testing on non-living subjects, and so forth."

"Well, not a bad solution," said the emperor's mother. "I just wish you didn't have to go and hurt him. He's not a bad boy."

"Sorry. I was desperate. I think we all were." Raising his voice, Nick added, "Desperate enough to edit this out for replay. Right?"

"Right!" chorused the auditorium, the people in the parking lot, the entire network viewing audience, and the emperor's mother, without hesitation.

Nick turned back to Sadie. "Come now, I think we'd better go before the traffic gets bad."

"Just one question, you two!" called a voice from a place near the center of the auditorium.

Nick and Sadie turned towards the audience, surprised. "Yes?" asked Nick.

A man in a bright yellow sweater stood up and looked down at the clipboard he was holding in his left hand. "Can you give us any idea *when* the next working model of completely transparent reality will be ready?"

Nick looked at Sadie and then at the again totally silent auditorium. He took off the rose headpiece and wiped the sweat off his forehead and cheeks. "We'll call you," he said and exited, with Sadie, stage right.

FERRYMAN

BY ERIC BROWN

Richard Lincoln sat in the darkened living room and half-listened to the radio news. More unrest in the East; riots and protests against the implantation process in India and Malaysia. The President of France had taken his life, another suicide statistic to add to the growing list.... The news finished and was followed by a weather report: more snow was forecast for that night and the following day.

Lincoln was hoping for a quiet shift when the bracelet around his wrist began to warm. He pushed himself from his armchair, crossed to the computer on the desk, and touched the bracelet to the screen.

The name and address of the deceased glowed in the darkness.

Despite the weather and the inconvenience of the late hour, as ever he felt the visceral thrill of embarkation, the anticipation of what was to come.

He memorised the address as he stepped into the hall and found his coat, already planning the route twenty miles over the moors to the dead man's town.

He was checking his pocket for the Range Rover's keys when he heard the muffled grumble, amplified by the snow, of a car's engine. His cottage was a mile from the nearest road, serviced by a potholed cart track. No one ever turned down the track by mistake, and he'd had no visitors in years.

He waited, as if half-expecting the noise to go away—but the vehicle's irritable whine increased as it fought through the snow and ice towards the cottage. Lincoln switched on the outside light and returned to the living room, pulling aside the curtain and peering out.

A white Fiat Panda lurched from pothole to pothole, headlights bouncing. It came to a stop outside the cottage, the sudden silence profound, and a second later someone climbed out.

Lincoln watched his daughter slam the door and pick her way carefully through the snow.

The door-bell chimed.

For a second he envisaged the tense confrontation that would follow, but the warm glow at his wrist gave him an excuse to reduce his contact with Susanne to a minimum.

He pulled open the door. She stood tall in an expensive white macintosh, collar turned up around her long, dark, snow-specked hair.

Her implant showed as a slight bulge at her temple.

She could hardly bring herself to look him in the eye. Which, he thought, was hardly surprising.

She gave a timid half-smile. "It's cold out here, Richard."

"Ah...Come in. This is a surprise. Why didn't you ring?"

"I couldn't talk over the phone. I needed to see you in person."

To explain herself, he thought; to excuse her recent conduct.

She swept past him, shaking the melted snow from her hair. She hung her coat in the hall and walked into the living room.

Lincoln paused behind her, his throat constricted with an emotion he found hard to identify. He knew he should have felt angry, but all he did feel was the desire for Susanne to leave.

"I'm sorry. I should have come sooner. I've been busy."

She was thirty, tall and good-looking and—*damn them*—treacherous genes had bequeathed her the unsettling appearance of her mother.

As he stared at her, Lincoln realised that he no longer knew the woman who was his daughter.

"But I'm here now," she said. "I've come about—"

He interrupted, his pulse racing. "I don't want to talk about your mother."

"Well I do," Susanne said. "This is important."

He recalled his excuse. "As a matter of fact it's impossible right now…" He held up his right hand, showing Susanne the band around his wrist.

"You've been called."

"It's quite a way—over the Pennines. Hebden Bridge. I should really be setting off. Look…make yourself at home. You know where the spare room is. We can…we'll talk in the morning, okay?"

He caught the flash of impatience on her face, soon doused by the realisation that nothing came between him and his calling.

She sighed. "Fine. See you in the morning."

Relief lifting from his shoulders like a weight, Lincoln nodded and hurried outside. Seconds later he was revving the Range Rover up the uneven track, into the darkness.

The road through the Pennines had been gritted earlier that night, and the snow that had fallen since had turned into a thin grey mush. Lincoln drove cautiously, his the only vehicle out this late. Insulated from the cold outside, he tried to forget about the presence of Susanne back at the cottage. He half-listened to a discussion programme on Radio Four. He imagined half-a-dozen dusty academics huddled in a tiny studio in Bush House. Cockburn, the Cambridge philosopher, had the microphone: *"It is indeed possible that individuals will experience a certain disaffection, even apathy, which is the result of knowing that there is more to existence than this life...."*

Lincoln wondered if this might explain the alienation he had felt for a year, since accepting his present position. But then he'd always had difficulty in showing his emotions, and consequently accepting that anyone else had emotions to show.

This life is a prelude, he thought, *a farce I've endured for fifty-five years—the end of which I look forward to with anticipation.*

It took him almost two and a half hours to reach Hebden Bridge. The small town, occupying the depths of a steep valley, was dank and quiet in the continuing snowfall. Streetlights sparkled through the darkness.

He drove through the town and up a steep hill, then turned right up an even steeper minor road. Hillcrest Farm occupied a bluff overlooking the acute incision of the valley. Coachlights burned orange around the front porch. A police car was parked outside.

Lincoln climbed from the Range Rover and hurried across to the porch. He stood for a second before pressing the doorbell, composing himself. He always found it best to adopt a neutral attitude until he could assess the mood of the bereaved family: more often than not the mood in the homes of the dead was one of excitement and anticipation.

Infrequently, especially if the bereaved were religious, a more formal grief prevailed.

He pressed the bell and seconds later a ruddy-faced local constable opened the door. "There you are. We've been wondering if you'd make it, weather like it is."

"Nice night for it," Lincoln said, stepping into the hall.

The constable gestured up a narrow flight of stairs. "The dead man's a farmer—silly bugger went out looking for a lost ewe. Heart attack. His daughter was out with him—but he was dead by the time she fetched help. He's in the front bedroom."

Lincoln followed the constable up the stairs and along a corridor. The entrance to the bedroom was impossibly low; both men had to stoop as if entering a cave.

He saw the bereaved family first, half-a-dozen men and women in their twenties and thirties, seated around the bed on dining chairs. An old woman, presumably the farmer's widow, sat on the bed itself, her husband's lifeless blue hand clutched in hers.

Lincoln registered the looks he received as he entered the room: the light of hope and gratitude burned in the eyes of the family, as if he, Lincoln himself, were responsible for what would happen over the course of the next six months.

The farmer lay fully dressed on the bed, rugged and grey like the carving of a knight on a sarcophagus.

An actor assuming a role, Lincoln nodded with suitable gravity to each of the family in turn.

"If anyone has any questions, anything at all, I'll be glad to answer them." It was a line he came out with every time to break the ice, but he was rarely questioned these days.

He stepped forward and touched his bracelet to the dead man's temple, where his implant raised a veined, weather-worn rectangle beneath the skin. The nanomeks would now begin the next stage of the process, the preparation of the body for its onward journey.

"I'll fetch the container," he said—he never called it a coffin—and nodded to the constable.

Together they carried the polycarbon container from the back of the Range Rover, easing it around the bends in the stairs. The family formed a silent huddle outside the bedroom door. Lincoln and the constable passed inside and closed the door behind them.

They lifted the corpse into the container and Lincoln sealed the sliding lid. The job of carrying the container down the stairs—attempting to maintain dignity in the face of impossible angles and improbable bends—was made all the more difficult by the presence of the family, watching from the stair landing.

Five minutes of gentle coaxing and patient lifting and turning, and the container was in the back of the Range Rover.

The constable handed over a sheaf of papers, which Lincoln duly signed and passed back. "I'll be on my way, Mr Lincoln," the constable said. "See you later." He waved and climbed into his squad car.

One of the farmer's daughters hurried from the house. "You'll stay for a cup of tea?"

Lincoln was about to refuse, then realised how cold he was. "Yes, that'd be nice. Thanks."

He followed her into a big, stone-flagged kitchen, an Aga stove filling the room with warmth.

He could tell that she had been crying. She was a plain woman in her mid-thirties, with the stolid, resigned appearance of the unfortunate sibling left at home to help with the farm work.

He saw the crucifix on a gold chain around her neck, and then noticed that her temple was without an implant. He began to regret accepting the offer of tea.

He sat at the big wooden table and wrapped his hands around the steaming mug. The woman sat down across from him, nervously meeting his eyes.

"It happened so quickly. I can hardly believe it. He had a weak heart—we knew that. We told him to slow down. But he didn't listen."

Lincoln gestured. "He was implanted," he said gently.

She nodded, eyes regarding her mug. "They all are, my mother, brothers and sisters." She glanced up at him, something like mute appeal in her eyes. "It seems that all the country is, these days."

When she looked away, Lincoln found his fingers straying to the outline of his own implant.

"But..." she whispered, "I'm sure things were...I don't know—*better* before. I mean, look at all the suicides—thousands of people every month take their lives..." She shook her head, confused. "Don't you think that people are less...less concerned now, less caring?"

"I've heard Cockburn's speeches. He says something along the same lines."

"I agree with him. To so many people this life is no longer so important. It's something to be got through, before what follows."

How could he tell her that he felt this himself?

He said, "But wasn't that what religious people thought about life, before the change?"

She stared at him as if he were an ignoramus. "No! Of course not. That might have been what atheists *thought* religious people felt.... But we love life, Mr Lincoln. We give thanks for the miracle of God's gift."

She turned her mug self-consciously between flattened palms. "I don't like what's happened to the world. I don't think it's *right*. I loved my father. We were close. I've never loved anyone quite so much." She looked up at him, her eyes silver with tears. "He was such a wonderful man. We attended church together. And then *they* came," she said with venom, "and everything changed. My father, he—" she could not stop the tears now— "he believed what they said. He left the Church. He had the implant, like all the rest of you."

He reached out and touched her hand. "Look, this might sound strange, coming from me, but I understand what you're saying. I might not agree, but I know what you're experiencing."

She looked at him, something like hope in her eyes. "You do? You really do? Then..." She fell silent, regarding the scrubbed pine table-top. "Mr Lincoln," she said at last, in a whispered entreaty, "do you really have to take him away?"

He sighed, pained. "Of course I do. It was his choice. He chose to be implanted. Don't you realise that to violate his trust, his choice..." He paused. "You said you loved him. In that case, respect his wishes."

She was slowly shaking her head. "But I love God even more," she said. "And I think that what is happening is wrong."

He drained his tea with a gesture of finality. "There'll be a religious service of your choice at the Station in two days' time," he said.

"And then...what then, Mr Lincoln?"

"Then he'll be taken, healed. In six months the process will be complete."

"Then he'll come back?"

"He'll be in contact before then, by recorded message, in around three months. Of course, he won't be able to travel until the six-month period has elapsed. Then he'll return."

"And after that?"

"It's his choice. Some choose to come back here and take up where they left off, resume their old lives. But sooner or later..." Lincoln shrugged. "In time he'll realise that there's more to life than what's here. Others prefer to make a clean break and work away from the start."

She said in a whisper, "What do they want with our dead, Mr Lincoln? Why are they doing this to us?"

He sighed. "You must have read the literature, seen the documentaries. It's all in there."

"But you...as a ferryman...surely you can tell me what they really want?"

"They want what they say—nothing more and nothing less."

A silence came between them. She was nodding, staring into her empty mug. He stood and touched her shoulder as he left the kitchen. He said goodbye to the family in the living room—gathered like the survivors of some natural catastrophe, unsure quite how to proceed— and let himself out through the front door.

He climbed into the Range Rover, turned and accelerated south towards the Onward Station.

He drove for the next hour through the darkness, high over the West Yorkshire moors, cocooned in the warmth of the vehicle with a symphony by Haydn playing counterpoint to the grumble of the engine.

Neither the music nor the concentration required to keep the vehicle on the road fully occupied his thoughts. The events at the farmhouse, and his conversation with the dead man's daughter, stirred memories and emotions he would rather not have recalled.

It was more than the woman's professed love for her dead father that troubled him, reminding him of his failed relationship with his Susanne. The fact that the farmer's daughter had foregone the implant stirred a deep anger within him. He had said nothing at the time, but now he wanted to return and plead with her to think again about undergoing the simple process that would grant her another life.

In the July of last year, at the height of summer, Lincoln's wife had finally left him. After thirty-five years of marriage she had walked out, moved to London to stay with Susanne until she found a place of her own.

In retrospect he was not surprised at her decision to leave; it was the inevitable culmination of years of neglect on his part. At the time, however, it had come as a shock—verification that the increasing disaffection he felt had at last destroyed their relationship.

He recalled their confrontation on that final morning as clearly as if it were yesterday.

Behind a barricade of suitcases piled in the hall, Barbara had stared at him with an expression little short of hatred. They had rehearsed the dialogue many times before.

"You've changed, Rich," she said accusingly. "Over the past few months, since taking the job."

He shook his head, tired of the same old argument. "I'm still the same person I always was."

She gave a bitter smile. "Oh, you've always been a cold and emotionless bastard, but since taking the job…"

He wondered if he had applied for the position because of who and what he was, a natural progression from the solitary profession of freelance editor of scholastic textbooks. Ferrymen were looked upon by the general public with a certain degree of wariness, much as undertakers had been in the past. They were seen as a profession apart.

Or, he wondered, had he become a ferryman to spite his wife?

There had been mixed reactions to the news of the implants and their consequences: many people were euphoric at the prospect of renewed life; others had been cautiously wary, not to say suspicious. Barbara had placed herself among the latter.

"There's no hurry," she had told Lincoln when he mentioned that he'd decided to have the operation. "I have no intention of dying, just yet."

At first he had taken her reluctance as no more than an affectation, a desire to be different from the herd. Most people they knew had had the implant: Barbara's abstention was a talking point.

Then it occurred to Lincoln that she had decided against having the implantation specifically to annoy him; she had adopted these frustrating affectations during the years of their marriage: silly things like refusing to holiday on the coast because of her dislike of the sea— or rather, because Lincoln loved the sea; deciding to become a vegetarian, and doing her damnedest to turn him into one, too.

Then, drunk one evening after a long session at the local, she had confessed that the reason she had refused the implant option was because she was petrified of what might happen to her after she died. She did not trust their motives.

"How…how do we know that they're telling the truth? How do we know what—what'll happen to us once *they* have us in their grasp?"

"You're making them sound like B-movie monsters," Lincoln said.

"Aren't they?"

He had gone through the government pamphlets with her, reiterated the arguments both for and against. He had tried to persuade her that the implants were the greatest advance in the history of humankind.

"But not everyone's going along with it," she had countered. "Look at all the protest groups. Look at what's happening around the world. The riots, political assassinations—"

"That's because they cling to their bloody superstitious religions," Lincoln had said. "Let's go over it again…"

But she had steadfastly refused to be convinced, and after a while he had given up trying to change her mind.

Then he'd applied to become a ferryman, and was accepted.

"I hope you feel pleased with yourself," Barbara said one day, gin-drunk and vindictive.

He had lowered his newspaper. "What do you mean?"

"I mean, why the hell do you want to work for them, do their dirty work?" Then she had smiled. "Because, Mr Bloody Ferryman, you'd rather side with them than with me. I'm only your bloody wife, after all."

And Lincoln had returned to the paper, wondering whether what she had said was true.

Over the next few weeks their relationship, never steady, had deteriorated rapidly. They lived separate lives, meeting for meals when, depending on how much she had drunk, Barbara could be sullenly uncommunicative or hysterically spiteful.

Complacent, Lincoln had assumed the rift would heal in time.

Her decision to leave had initially shocked him. Then, as the decision turned from threat to reality, he saw the logic of their separation—it was, after all, the last step in the process of isolation he had been moving towards for a long, long time.

He had pleaded with her, before she left, to think again about having the implant operation.

"The first resurrectees will be returning soon," he told her. "Then you'll find that you've nothing to fear."

But Barbara had merely shaken her head and walked out of his life.

He wrote to her at Susanne's address over the next couple of months, self-conscious letters expressing his hopes that Barbara was doing okay, would think again about having an implant. Reading the letters back to himself, he had realised how little he had said—how little there was to say—about himself and his own life.

Then last autumn, Lincoln had received a phone call from Susanne. The sound of her voice—the novelty of her call—told Lincoln that something was wrong.

"It's your mother—" he began.

"Dad…I'm sorry. She didn't want you to know. She was ill for a month—she wasn't in pain."

All he could say was, "What?" as a cold hollow expanded inside his chest.

"Cancer. It was inoperable."

Silence—then, against his better judgement, he asked, "Did...did she have the implant, Susanne?"

An even longer silence greeted the question, and Lincoln knew full well the answer.

"She didn't want a funeral," Susanne said. "I scattered her ashes on the pond at Rochester."

A week later he had travelled down to London. He called at his daughter's flat, but she was either out or ignoring him. He drove on to Rochester, his wife's birthplace, not really knowing why he was going but aware that, somehow, the pilgrimage was necessary.

He had stood beside the pond, staring into the water and weeping quietly to himself. Christ, he had hated the bitch at times—but, again, at certain times with Barbara he had also experienced all the love he had ever known.

As if to mock the fact of his wife's death, her immutable nonexistence, the rearing crystal obelisk of this sector's Onward Station towered over the town like a monument to humankind's new-found immortality, or an epitaph to the legion of dead and gone.

He had returned home and resumed his work, and over the months the pain had become bearable. His daughter's return, last night, had reopened the old wound.

A silver dawn was breaking over the horizon, revealing a landscape redesigned, seemingly inflated, by the night's snowfall. The Onward Station appeared on the skyline, a fabulous tower of spun glass scintillating in the light of the rising sun.

He visited the Station perhaps four or five times a week, and never failed to stare in awe—struck not only by the structure's ethereal architecture, but by what it meant for the future of humankind.

He braked in the car park alongside the vehicles of the dozen other ferrymen on duty today. He climbed out and pulled the polycarbon container from the back of the Range Rover, the collapsible chromium trolley taking its weight. His breath pluming before him in the ice-cold air, he hurried towards the entrance set into the sloping glass walls.

The interior design of the Station was arctic in its antiseptic inhospitality, the corridors shining with sourceless, polar light. At these times, as he maneuvered the trolley down the seemingly endless corridors, he felt that he was, truly, trespassing on territory forever alien.

He arrived at the preparation room and eased the container onto the circular reception table, opening the lid. The farmer lay unmoving in death, maintained by the host of alien nanomeks that later, augmented by others more powerful, would begin the resurrection process. They would not only restore him to life, strip away the years, but make him fit and strong again: the man who returned to Earth in six months would be physically in his thirties, but effectively immortal.

In this room, Lincoln never ceased to be overcome by the wonder, as might a believer at the altar of some mighty cathedral.

He backed out, pulling the trolley after him, and retraced his steps. To either side of the foyer, cleaners vacuumed carpets and arranged sprays of flowers in the Greeting rooms, ready to receive the day's returnees, their relatives and loved ones.

He emerged into the ice-cold dawn and hurried across to the Range Rover. On the road that climbed the hill behind the Station, he braked and sat for ten minutes staring down at the diaphanous structure.

Every day a dozen bodies were beamed from this Station to the lightship in geo-sync orbit, pulses of energy invisible during the daylight hours. At night the pulses were blinding columns of blue lightning, illuminating the land for miles around.

From Earth orbit, the ships phased into trans-c mode and reached the aliens' home planet in days. There the dead were revived, brought back to life and gradual consciousness by techniques of medical science that experts on Earth were still trying to comprehend. After six months of rehabilitation and instruction, the resurrected had the choice of returning to Earth, or beginning their missions immediately. Children and youths under the age of twenty were returned, to live their lives until adulthood and such time as they decided to progress onwards.

Lincoln looked up, into the rapidly fading darkness. A few bright stars still glimmered, stars that for so long had been mysterious and unattainable—and now, hard though it was sometimes to believe, had been thrown open to humankind by the beneficence of beings still mistrusted by many, but accepted by others as saviours.

And why had the aliens made their offer to humankind?

There were millions upon millions of galaxies out there, the aliens said, billions of solar systems, and countless, literally countless, planets that sustained life of various kinds. Explorers were needed, envoys and ambassadors, to discover new life, and make contact, and spread the greetings of the civilised universe far and wide.

Lincoln stared up at the fading stars and thought what a wondrous fact, what a miracle it was; he considered the new worlds out there, waiting to be discovered, strange planets and civilisations, and it was almost too much to comprehend that, when he died and was reborn, he too would venture out on that greatest diaspora of all.

He drove home slowly, tired after the exertions of the night. Only when he turned down the cart track, and saw the white Fiat parked outside the cottage, was he reminded of his daughter.

He told himself that he would make an effort today: he would not reprimand her for saying nothing about Barbara's illness, wouldn't even question her. God knows, he had never done anything to earn her trust and affection: it was perfectly understandable that she had complied with her mother's last wishes.

Still, despite his resolve, he felt a slow fuse of anger burning within him as he climbed from the Range Rover and let himself into the house.

He moved to the kitchen to make himself a coffee, and as he was crossing the hall he noticed that Susanne's coat was missing from the stand, and likewise her boots from beneath it.

From the kitchen window he looked up at the broad sweep of the moorland, fleeced in brilliant snow, to the gold- and silver-laminated sunrise.

He made out Susanne's slim figure silhouetted against the brightness. She looked small and vulnerable, set against such vastness, and Lincoln felt something move within him, an emotion like sadness and regret, the realisation of squandered opportunity.

On impulse he fetched his coat, left the cottage and followed the trail of her deep footprints up the hillside to the crest of the rise.

She heard the crunch of his approach, turned and gave a wan half-smile. "Admiring the view," she whispered.

He stood beside her, staring down at the limitless expanse of the land, comprehensively white save for the lee sides of the dry-stone walls, the occasional distant farmhouse.

Years ago he had taken long walks with Susanne, enjoyed summer afternoons together on the wild and undulating moorland. Then she had grown, metamorphosed into a teenager he had no hope of comprehending, a unique individual—no longer a malleable child—over whom he had no control. He had found himself, as she came more and more to resemble her mother and take her side in every argument, in a minority of one.

He had become increasingly embittered, over the years. Now he wanted to reach out to Susanne, make some gesture to show her that he cared, but found himself unable even to contemplate the overture of reconciliation.

In the distance, miles away on the far horizon, was the faerie structure of the Station, its tower flashing sunlight.

At last she said, "I'm sorry," so softly that he hardly heard.

His voice seemed too loud by comparison. "I understand," he said.

She shook her head. "I don't think you do." She paused. Tears filled her eyes, and he wondered why she was crying like this.

"Susanne…"

"But you *don't* understand."

"I do," he said gently. "Your mother didn't want me to know about her illness—she didn't want me around. Christ, I was a pain enough to her when she was perfectly well."

"It wasn't that," Susanne said in a small voice. "You see, she didn't want you to know that she'd been wrong."

"Wrong?" He stared at her, not comprehending. "Wrong about what?"

She took a breath, said, "Wrong about the implant," and tears escaped her eyes and tracked down her cheeks.

Lincoln felt something tighten within his chest, constrict his throat, making words difficult.

"What do you mean?" he asked at last.

"Faced with death, in the last weeks…it was too much. I…I persuaded her to think again. At last she realised she'd been wrong.

A week before she died, she had the implant." Susanne looked away, not wanting, or not daring, to look upon his reaction to her duplicity.

He found it impossible to speak, much less order his thoughts, as the realisation coursed through him.

Good God. *Barbara…*

He felt then love and hate, desire and a flare of anger.

Susanne said: "She made me swear not to tell you. She hated you, towards the end."

"It was my fault," he said. "I was a bastard. I deserved everything. It's complex, Susanne, so bloody damned complex—loving someone and hating them at the same time, needing to be alone and yet needing what they can give."

A wind sprang up, lifting a tress of his daughter's hair. She fingered it back into place behind her ear. "I heard from her three months ago—a kind of CD thing delivered from my local Station. She told me that she'd been terribly cruel in not telling you. I…I meant to come up and tell you earlier, but I had no idea how you'd react. I kept putting it off. I came up yesterday because it was the last chance before she returns."

"When?" Lincoln asked, suddenly aware of the steady pounding of his heart.

"Today," Susanne said. She glanced at her watch. "At noon today—at this Station."

"This Station?" Lincoln said. "Of all the hundreds in Britain?" He shook his head, some unnameable emotion making words difficult. "What…what does she want?"

"To see you, of course. She wants to apologise. She told me she's learned a great many things up there, and one of them was compassion."

Oh, Christ, he thought.

"Susanne," he said, "I don't think I could face your mother right now."

She turned to him. "Please," she said, "please, this time, can't you make the effort—for me? What do you think it's been like, watching you two fight over the years?"

Lincoln balked at the idea of meeting this resurrected Barbara, this reconstructed, compassionate creature. He wanted nothing of her pity.

"Look," Susanne said at last, "she's leaving soon, going to some star I can't even pronounce. She wants to say goodbye."

Lincoln looked towards the horizon, at the coruscating tower of the Station.

"We used to walk a lot round here when I was young," Susanne said. There was a note of desperation in her voice, a final appeal.

Lincoln looked at his watch. It was almost nine. They could just make it to the Station by midday, if they set off now.

He wondered if he would have been able to face Barbara, had she intended to stay on Earth.

At last, Lincoln reached out and took his daughter's hand.

They walked down the hill, through the snow, towards the achingly beautiful tower of the Onward Station.

GREAT WESTERN

BY KIM NEWMAN

Cleared paths were no good for Allie. She wasn't supposed to be after rabbits on Squire Maskell's land. Most of Alder Hill was wildwood, trees webbed together by a growth of bramble nastier than barbwire. Thorns jabbed into skin and stayed, like bee-stingers.

Just after dawn, the air had a chilly bite but the sunlight was pure and strong. Later, it would get warm; now, her hands and knees were frozen from dew-damp grass and iron-hard ground.

The Reeve was making a show of being tough on poaching, handing down short, sharp sentences. She'd already got a stripe across her palm for setting snares. Everyone west of Bristol knew Reeve Draper was Maskell's creature. Serfdom might have been abolished, but the old squires clung to their pre-War position, through habit as much as tenacity.

Since taking her lash, administered under the village oak by Constable Erskine with a razor-strop, she'd grown craftier. Wiry enough to tunnel through bramble, she made and travelled her own secret, thorny paths. She'd take Maskell's rabbits, even if the Reeve's Constable striped her like a tiger.

She set a few snares in obvious spots, where Stan Budge would find and destroy them. Maskell's gamekeeper wouldn't be happy if he thought no one was even trying to poach. The trick was to set snares invisibly, in places Budge was too grown-up, too far off the ground, to look.

Even so, none of her nooses had caught anything.

All spring, she'd been hearing gunfire from Alder Hill, resonating across the moors like thunder. Maskell had the Gilpin brothers out with Browning rifles. They were supposed to be ratting, but the object of the exercise was to end poaching by killing off all the game.

There were rabbit and pigeon carcasses about, some crackly bone bundles in packets of dry skin, some recent enough to seem shocked to death. It was a sinful waste, what with hungry people queueing up for parish hand-outs. Quite a few trees had yellow-orange badges, where Terry or Teddy Gilpin had shot wide of the mark. Squire Maskell would not be heartbroken if one of those wild shots finished up in her.

Susan told her over and over to be mindful of men with guns. She had a quite reasonable horror of firearms. Too many people on Sedgmoor died with their gumboots on and a bullet in them. Allie's Dad and Susan's husband, for two. Susan wouldn't have a gun in the house.

For poaching, Allie didn't like guns anyway. Too loud. She had a catapult made from a garden fork, double-strength rubber stretched between steel tines. She could put a nail through a half-inch of plywood from twenty-five feet.

She wriggled out of her tunnel, pushing aside a circle of bramble she'd fixed to hinge like a lid, and emerged in a clearing of loose earth and shale. During the Civil War, a bomb had fallen here and fizzled. Eventually, the woods would close over the scar.

When she stood up, she could see across the moors, as far as Achelzoy. At night, the infernal lights of Bridgwater pinked the horizon, clawing a ragged red edge in the curtain of dark. Now, she could make out the road winding through the wetlands. The sun, still low, glinted and glimmered in sodden fields, mirror-fragments strewn in a carpet of grass. There were dangerous marshes out there. Cows were sucked under if they set a hoof wrong.

Something moved near the edge of the clearing.

Allie had her catapult primed, her eye fixed on the rabbit. Crouching, still as a statue, she concentrated. Jack Coney nibbled on nothing, unconcerned. She pinched the nailhead, imagining a point between the ears where she would strike.

A noise sounded out on the moor road. The rabbit vanished, startled by the unfamiliar rasp of an engine.

"'S'blood," she swore.

She stood up, easing off on her catapult. She looked out towards Achelzoy. A fast-moving shape was coming across the moor.

The rabbit was lost. Maskell's men would soon be about, making the woods dangerous. She chanced a maintained path and ran swiftly downhill. At the edge of Maskell's property, she came to a stile and vaulted it—wrenching her shoulder, but no matter—landing like a cat on safe territory. Without a look back at the "TRESPASSERS WILL BE VENTILATED" sign, she traipsed between two rows of trees, towards the road.

The path came out half a mile beyond the village, at a sharp kink in the moor road. She squatted with her back to a signpost, running fingers through her hair to rid herself of tangles and snaps of thorn.

The engine noise was nearer and louder. She considered putting a nail in the nuisance-maker's petrol tank to pay him back for the rabbit. That was silly. Whoever it was didn't know what he'd done.

She saw the stranger was straddling a Norton. He had slowed to cope with the winds of the moor road. Every month, someone piled up in one of the ditches because he took a bend too fast.

To Allie's surprise, the motorcyclist stopped by her. He shifted goggles up to the brim of his hat. He looked as if he had an extra set of eyes in his forehead.

There were care-lines about his eyes and mouth. She judged him a little older than Susan. His hair needed cutting. He wore leather trews, a padded waistcoat over a dusty khaki shirt, and gauntlets. A brace of pistols was holstered at his hips, and he had a rifle slung on the Norton, within easy reach.

He reached into his waistcoat for a pouch and fixings. Pulling the drawstring with his teeth, he tapped tobacco onto a paper and rolled himself a cigarette one-handed. It was a clever trick, and he knew it. He stuck the fag in his grin and fished for a box of Bryant and May.

"Alder," he said, reading from the signpost. "Is that a village?"

"Might be."

"Might it?"

He struck a light on his thumbnail and drew a lungful of smoke, held in for a moment like a hippie sucking a joint, and let it funnel out through his nostrils in dragon-plumes.

"Might it indeed?"

He didn't speak like a yokel. He sounded like a wireless announcer, maybe even more clipped and starched.

"If, hypothetically, Alder were a village, would there be a hostelry there where one might buy breakfast?"

"Valiant Soldier don't open till lunchtime."

The Valiant Soldier was Alder's pub, and another of Squire
Maskell's businesses.

"Pity."

"How much you'm pay for breakfast?" she asked.

"That would depend on the breakfast."

"Ten bob?"

The stranger shrugged.

"Susan'll breakfast you for ten bob."

"Your mother."

"No."

"Where could one find this Susan?"

"Gosmore Farm. Other end of village."

"Why don't you get up behind me and show me where to go?"

She wasn't sure. The stranger shifted forward on his seat, making
space.

"I'm Lytton," the stranger said.

"Allie," she replied, straddling the pillion.

"Hold on tight."

She took a grip on his waistcoat, wrists resting on the stocks of his
guns.

Lytton pulled down his goggles and revved. The bike sped off. Allie's
hair blew into her face and streamed behind her. She held tighter,
pressing against his back to keep her face out of the wind.

When they arrived, Susan had finished milking. Allie saw her
washing her hands under the pump by the back door.

Gosmore Farm was a tiny enclave circled by Maskell's land. He
had once tried to get the farm by asking the newly widowed Susan

to marry him. Allie couldn't believe he'd actually thought she might consent. Apparently, Maskell didn't consider Susan might hold a grudge after her husband's death. He now had a porcelain doll named Sue-Clare in the Manor House, and a pair of terrifying children.

Susan looked up when she heard the Norton. Her face was set hard. Strangers with guns were not her favourite type of folk.

Lytton halted the motorcycle. Allie, bones shaken, dismounted, showing herself.

"He'm pay for breakfast," she said. "Ten bob."

Susan looked the stranger over, starting at his boots, stopping at his hips.

"He'll have to get rid of those filthy things."

Lytton, who had his goggles off again, was puzzled.

"Guns, she means," Allie explained.

"I know you feel naked without them," Susan said sharply. "Unmanned, even. *Magna Carta* rules that no Englishman shall be restrained from bearing arms. It's that fundamental right which keeps us free."

"That's certainly an argument," Lytton said.

"If you want breakfast, yield your fundamental right before you step inside my house."

"That's a stronger one," he said.

Lytton pulled off his gauntlets and dropped them into the pannier of the Norton. His fingers were stiff on the buckle of his gunbelt, as if he had been wearing it for many years until it had grown into him like a wedding ring. He loosened the belt and held it up.

Allie stepped forward to take the guns.

"Allison, no," Susan insisted.

Lytton laid the guns in the pannier and latched the lid.

"You have me defenceless," he told Susan, spreading his arms.

Susan squelched a smile and opened the back door. Kitchen smells wafted.

A good thing about Lytton's appearance at Gosmore Farm was that he stopped Susan giving Allie a hard time about being up and about before dawn. Susan had no illusions about what she did in the woods.

Susan let Lytton past her into the kitchen. Allie trotted up.

"Let me see your hands," Susan said.

Allie showed them palms down. Susan noted dirt under nails and a few new scratches. When Allie showed her palms, Susan drew a fingernail across the red strop-mark.

"Take care, Allie."

"Yes'm."

Susan hugged Allie briefly, and pulled her into the kitchen.

Lytton had taken a seat at the kitchen table and was loosening his heavy boots. Susan had the wireless on, tuned to the Light Programme. Mark Radcliffe introduced the new song from Jarvis Cocker and His Wurzels, "The Streets of Stogumber.' A frying pan was heating on the cooker, tiny trails rising from the fat.

"Allie, cut our guest some bacon."

"The name's Lytton."

"I'm Susan Ames. This is Allison Conway. To answer your unasked question, I'm a widow, she's an orphan. We run this farm ourselves."

"A hard row to plough."

"We're still above ground."

Allie carved slices off a cured hock that hung by the cooker. Susan took eggs from a basket, cracked them into the pan.

"Earl Gray or Darjeeling?" Susan asked Lytton.

"The Earl."

"Get the kettle on, girl," Susan told her. "And stop staring."

Allie couldn't remember Susan cooking for a man since Mr Ames was killed. It was jarring to have this big male, whiffy from the road and petrol, invading their kitchen. But also a little exciting.

Susan flipped bacon rashers, busying herself at the cooker. Allie filled the kettle from the tap at the big basin.

"Soldier, were you?" Susan asked Lytton, indicating his shoulder. There was a lighter patch on his shirtsleeve where rank insignia had been cut away. He'd worn several pips.

The stranger shrugged.

"Which brand of idiot?"

"I fought for the southeast."

"I'd keep quiet about that if you intend to drink in The Valiant Soldier."

"I'd imagined Wessex was mostly neutral."

"Feudal order worked perfectly well for a thousand years. It wasn't just landed gentry who resisted London Reforms. There are plenty of jobless ex-serfs around, nostalgic for their shackles and three hot meals a day."

"Just because it lasted a long time doesn't mean it was a good thing."

"No argument from me there."

"Mr Ames was a Reformist too," Allie said.

"Mr Ames?"

"My late husband. He opened his mouth too much. Some loyal retainers shut it for him."

"I'm sorry."

"Not your problem."

Susan wasn't comfortable talking about her husband. Mr Ames had been as much lawyer as farmer, enthusiastically heading the Sedgmoor

District Committee during the Reconstruction. He didn't realise it took more than a decision made in London Parliament to change things in the West. London was a long way off.

Allie brought Susan plates. Susan slid bacon and eggs from the pan.

"Fetch the tomato chutney from the preserves shelf," she said.

Outside, someone clanged the bell by the gate. Lytton's hand slipped quietly to his hip, closing where the handle of a revolver would have been.

Susan looked at the hot food on the table, and frowned at the door.

"Not a convenient time to come visiting," she said.

Hanging back behind Susan, Allie still saw who was in the drive. Constable Erskine was by the bell, vigorously hammering with the butt of his police revolver. His blue knob-end helmet gave him extra height. His gun-belt was in matching blue. Reeve Draper, arms folded, cringed at the racket his subordinate was making. Behind the officers stood Terry and Teddy Gilpin, Browning rifles casually in their hands, long coats brushing the ground.

"Goodwife Ames," shouted the Reeve. "This be a court order."

"Leave your guns."

"Come you now, Goodwife Ames. By right of law…"

Erskine was still clanging. The bell came off its hook and thunked on the ground. The Constable shrugged a grin and didn't holster his pistol.

"I won't have guns on my property."

"Then come and be served. This yere paper pertains to your cattle. The decision been telegraphed from Taunton Magistrates. You'm to surrender all livestock within thirty days, for slaughter. It be a safety measure."

Susan had been expecting something like this.

"There are no mad cows on Gosmore Farm."

"Susan, don't be difficult."

"It's Mrs Ames, Mr Reeve Draper."

The Reeve held up a fawn envelope.

"You'm know this has to be done."

"Will you be slaughtering Maskell's stock?"

"He took proper precautions, Susan. Can't be blamed. He'm been organic since 'fore the War."

Susan snorted a laugh. Everyone knew there'd been mad cow disease in the Squire's herd. He'd paid off the inspectors and rendered the affected animals into fertiliser. It was Susan who'd never used infected feed, never had a sick cow. This wasn't about British beef; this was about squeezing Gosmore Farm.

"Clear off," Allie shouted.

"Poacher girl," Erskine sneered. "Lookin' for a matchin' stripe on your left hand?"

Susan turned on the Constable.

"Don't you threaten Allison. She's not a serf."

"Once a serf, always a serf."

"What are they here for?" Susan nodded to the Gilpin brothers. "D'you need two extra guns to deliver a letter?"

Draper looked nervously at the brothers. Terry, heavier and nastier, curled his fingers about the trigger guard of his Browning.

"Why didn't Maskell come himself?"

Draper carefully put the letter on the ground, laying a stone on top of it.

"I'll leave this here, Goodwife. You'm been served with this notice."

Susan strode towards the letter.

Terry hawked a stream of spit, which hit the stone and splattered the envelope. He showed off his missing front teeth in an idiot leer.

Draper was embarrassed and angered, Erskine delighted and itchy.

"My sentiments exactly, Goodman Gilpin," said Susan. She kicked the stone and let the letter skip away in the breeze.

"Mustn't show disrespect for the law," Erskine snarled. He was holding his gun rightway round, thumb on the cock-lever, finger on the trigger.

From the kitchen doorway, close behind Allie, Lytton said, "Whose law?"

Allie stepped aside and Lytton strode into the yard. The four unwelcome visitors looked at him.

"Widow Ames got a stay-over guest," Erskine said, nastily.

"B'ain't no business of yourn, Goodman," said the Reeve to Lytton.

"And what if I make it my business?"

"You'm rue it."

Lytton kept his gaze steady on the Reeve, who flinched and blinked.

"He hasn't got a gun," Susan said, voice betraying annoyance with Lytton as much as with Maskell's men. "So you can't have a fair fight."

Mr Ames had been carrying a Webley when he was shot. The magistrate, Sue-Clare Maskell's father, ruled it a fair fight, exonerating on the grounds of self-defence the Maskell retainer who'd killed Susan's husband.

"He'm interfering with due process, Mr Reeve," Erskine told Draper. "We could detain him for questioning."

"I don't think that'll be necessary," Lytton said. "I just stopped at Gosmore Farm for bacon and eggs. I take it there's no local ordinance against that."

"Goodwife Ames don't have no bed and breakfast license," Draper said.

"Specially *bed*," Erskine added, leering.

Lytton strolled casually towards his Norton. And his guns.

"Maybe I should press on. I'd like to be in Dorset by lunchtime."

Terry's rifle was fixed on Lytton's belly, and swung in an arc as Lytton walked. Erskine thumb-cocked his revolver, ineptly covering the sound with a cough.

"Tell Maurice Maskell you've delivered your damned message," Susan said, trying to get between Lytton and the visitors' guns. "And tell him he'll have to come personally next time."

"You'm stay away from thic rifle, Goodman," the Reeve said to Lytton.

"Just getting my gloves," Lytton replied, moving his hands away from the holstered rifle towards the pannier where his pistols were.

Allie backed away towards the house, stomach knotted.

"What's she afraid of?" Erskine asked, nodding at her.

"Don't touch thic fuckin' bike," Terry shouted.

Allie heard the guns going off, louder than rook-scarers. An apple-sized chunk of stone exploded on the wall nearby, spitting chips in her face. The fireflashes were faint in the morning sun, but the reports were thunderclaps.

Erskine had shot, and Terry. Lytton had slipped down behind his motorcycle, which had fallen on him. There was a bright red splash of blood on the ground. Teddy was bringing up his rifle.

She scooped a stone and drew back the rubber of her catapult.

Susan screamed for everyone to stop.

Allie loosed the stone and raised a bloody welt on Erskine's cheek.

Susan slapped Allie hard and hugged her. Erskine, arm trembling with rage, blood dribbling on his face, took aim at them. Draper put a hand on the Constable's arm, and forced him to holster his gun. At a nod from the Reeve, Teddy Gilpin took a look at Lytton's wound and reported that it wasn't serious.

"This be bad, Goodwife Ames. It'd not tell well for you if'n it came up at magistrate's court. We'm be back on Saturday, with the vet. Have your animals together so they can be destroyed."

He walked to his police car, his men loping after him like dogs. Terry laughed a comment to Erskine about Lytton.

Allie impotently twanged her catapult at them.

"Help me get this off him," Susan said.

The Norton was a heavy machine, but between them they hefted it up. The pannier was still latched down. Lytton had not got to his guns. He lay face-up, a bright splash of red on his left upper arm. He was gritting his teeth against the hurt, shaking as if soaked to the skin in ice-water.

Allie didn't think he was badly shot. Compared to some.

"You stupid man," Susan said, kicking Lytton in the ribs. "You stupid, stupid man!"

Lytton gulped in pain and cried out.

It wasn't as if they had much livestock. Allie looked round at the eight cows, all with names and personalities, all free of the madness. Gosmore Farm had a chicken coop, a vegetable garden, a copse of apple trees and a wedge of hillside given over to grazing. It was a struggle to eke a living; without the milk quota, it would be hopeless.

It was wrong to kill the cows.

Despair lodged like a stone in Allie's heart. This was not what the West should be. When younger, she'd read Thomas Hardy's Wessex novels, *The Sheriff of Casterbridge* and *Under the Hanging Tree*, and she still followed *The Archers*. In storybook Wessex, men like Squire Maskell always lost. Alder needed Dan Archer, the wireless hero, to stride into The Valiant Soldier, six-guns blazing, and lay the vermin in the dirt.

There was no Dan Archer.

Susan held all her rage in, refusing to talk about the cows and Maskell. She always concentrated on what she called "the job at hand." Just now, she was nursing Lytton. Erskine's shot had gone right through his arm. Allie had looked for but not found the bullet, to give him as a souvenir. He'd lost blood, but he would live.

Allie hugged Pansy, her favourite, and brushed flies away from the cow's gummy eyes.

"I won't let they hurt you," she vowed.

But what could she do?

Depressed, she trudged down to the house.

Lytton was sitting up on the cot in the living room, with his shirt off and a clean white bandage tight around his arm. Allie saw he had older scars. This was not the first time he'd been shot. He was sipping a mug of hot tea. Susan, bustling furiously, tidied up around him. When he saw Allie, Lytton smiled.

"Susan's been telling me about this Maskell character. He seems to like to have things his way."

The door opened and Squire Maskell stepped in.

"That I do, sir."

He was dressed for church, in a dark suit and kipper tie. He knew enough not to wear a gunbelt on Gosmore Farm, though Allie guessed he was carrying a small pistol in his armpit. He had shot Allie's Dad with such a gun, in a dispute over wages. Allie barely remembered her father, who had been indentured on Maskell's farm before the War and an NFU rep afterwards.

"I don't remember extending an invitation, Squire," Susan said evenly.

"Susan, Susan, things could be so much more pleasant between us. We are neighbours."

"In the same way a pack of dogs are neighbours to a fox gone to earth."

Maskell laughed without humour.

"I've come to extend an offer of help."

Susan snorted. Lytton said nothing but looked Maskell over with eyes that saw the gun under the hankie-pocket and the knife in the boot.

"I understand you have BSE problems? My condolences."

"There's no mad cow disease in my herd."

"It's hardly a herd, Susan. It's a gaggle. But without them, where would you be?"

Maskell spread empty hands.

"This place is hardly worth the upkeep, Susan. You're only sticking at it because you have a nasty case of Stubborn Fever. The land is worthless to anyone but me. Gosmore Farm is a wedge in my own holdings. It would be so convenient if I could take down your fences, if I could incorporate your few acres into the Maskell farm."

"Now tell me something I don't know."

"I can either buy from you now above the market value, or wait a while and buy from the bank at a knock-down price. I'm making an offer now purely out of neighbourly charity. The old ways may have changed, but as Squire I still feel an obligation to all who live within my bailiwick."

"The only obligation your forefathers felt was to sweat the serfs into early graves and beget illegitimate cretins on terrorised girls. Have you noticed how the Maskell chin shows up on those Gilpin creatures?"

Maskell was angry now, but trying to keep calm. A vein throbbed by his eye.

"Susan, you're upset, I see that. But you must be realistic. Despite what you think, I don't want to see you on the mercy of the parish. Robert Ames was a good friend to me, and…"

"You can fuck off, Maskell," Susan spat. "Fuck *right* off."

The Squire's smile drained away. He was close to sputtering. His Maskell chin wobbled.

"Don't ever mention my husband again. And now leave."

"Susan," he pleaded.

"I think Goodwife Ames made herself understood," Lytton said.

Maskell looked at the wounded man. Lytton eased himself gingerly off the cot, expanding his chest, and stood. He was tall enough to have to bow his head under the beamed ceiling.

"I don't believe I've had…"

"Lytton," he introduced himself.

"And you would be… ?"

"I would be grateful if you left the house as Goodwife Ames wishes. And fasten the gate on your way out. There's a Country Code, you know."

"Good day," Maskell said, not meaning it, and left.

There was a moment of silence.

"That's the second time you've taken it on yourself to act for me," Susan said, angrily. "Have I asked your help?"

Lytton smiled. His hard look faded and he seemed almost mischievous.

"I beg pardon, Goodwife."

"Don't do it again, Lytton."

By the next day, Lytton was well enough to walk. But he couldn't ride: if he tried to grip the Norton's left handlebar, it was as if a red-hot poker were pressed to his bicep. They were stuck with him.

"You can do odd jobs for your keep," Susan allowed. "Allie will show you how."

"Can he come feed the chickens?" Allie asked, excited despite herself. "I can get the eggs."

"That'll be a start."

Susan walked across to the stone sheds where the cows spent the night, to do the milking. Allie took Lytton by the hand and led him round to the chicken coop.

"Maskell keeps his chickens in a gurt prison," Allie told him. "Clips their beaks with pliers, packs they in alive like sardines. If one dies, t'others eat her. They'm *cannibal* chickens..."

They turned round the corner.

The chicken coop was silent. Tears pricked the backs of Allie's eyes. Lumps of feathery matter lay in the scarlet-stained straw.

Her first thought was that a fox had got in.

Lytton lifted up a flap of chicken wire. It had been cut cleanly.

The coop was a lean-to, a chickenwire frame built against the house. On the stone wall was daubed a sign in blood, an upside-down tricorn fork in a circle.

"Travellers," Allie spat.

There was a big Gypsy Site at Glastonbury. Since the War, Travellers were supposed to stay on the sites, living off the dole. But they were called Travellers because they didn't like to keep to one place. They were always escaping from sites and raiding farms and villages.

Lytton shook his head.

"Hippies are hungry. They'd never have killed and left the chickens. And smashed the eggs."

The eggs had been gathered and carefully stamped on.

"Some hippies be veggie."

The blood was still fresh. Allie didn't see how this could have been done while they were asleep. The killers must have struck fast, or the chickens would have squawked.

"Where's your vegetable garden?" Lytton asked.

Allie's heart pounded like a fist.

She showed him the path to the garden, which was separated from the orchard by a thick hedge. Beanpoles had been wrenched from the earth and used to batter and gouge the rest of the crops. Cabbages were squashed, young carrots pulped by boot heels, marrows exploded. The greenhouse was a skeleton, every pane of glass broken, tomato plants strewn and flattened inside. Even the tiny herb patch Allie had been given for herself was dug up and scattered.

Allie sobbed. Liquid squirted from her eyes and nose. Hundreds of hours of work destroyed.

There was a twist of cloth on the frame of the greenhouse. Lytton examined it: a tie-dyed poncho, dotted with emblem badges of marijuana leaves, multi-coloured swirls and cartoon cats.

"Hippies," Allie yelled. "Fuckin' hippies."

Susan appeared at the gate. She swayed, almost in a swoon, and held the gate to stay standing.

"Hippies didn't do this," Lytton said.

He lifted a broken tomato plant from the paved area by the greenhouse door and pointed at a splashed yellow stain.

"Allie, where've you seen something like this recently?"

It came to her.

"Terry Gilpin. When he spat at thic letter."

"He has better aim with his mouth than his gun," Lytton commented, wincing. "Thankfully."

Lytton stood by his Norton, lifting his gauntlets out of the pannier.

"Are you leaving?" Allie asked.

"No," Lytton said, taking his gunbelt, "I'm going down to the pub."

He settled the guns on his hips and fastened the buckle. The belt seemed to give him strength, to make him stand straighter.

Susan, still shocked, didn't protest.

"Are you'm going to shoot Squire Maskell?" Allie asked.

That snapped Susan out of it. She took Allie and shook her by the shoulders, keening wordlessly.

"I'm just going to have a lunchtime drink."

Allie hugged Susan fiercely. They were on the point of losing everything, but gave each other the last of their strength. There was something Maskell couldn't touch.

Lytton strolled towards the front gate.

Allie pulled away from Susan. For a moment, Susan wouldn't let her go. Then, without words, she gave her blessing. Allie knew she was to look after Lytton.

He was halfway down the street, passing the bus shelter, disused since the service was cut, when Allie caught up with him. At the fork in the road where the village oak stood was The Valiant Soldier.

They walked on.

"I hope you do shoot him," she said.

"I just want to find out why he's so obsessed with Gosmore Farm, Allie. Men like Maskell always have reasons. That's why they're pathetic. You should only be afraid of men without reasons."

Lytton pushed open the door, and stepped into the public bar. This early, there were few drinkers. Danny Keogh sat in his usual seat, wooden leg unslung on the floor beside him. Teddy Gilpin was swearing at the Trivial Pursuit machine, and his brother was nursing a half of scrumpy and a packet of crisps, ogling the Tiller Girl in UI.

Behind the bar, Janet Speke admired her piled-up hair in the long mirror. She saw Lytton and displayed immediate interest, squirming tightly in an odd way Allie almost understood.

Terry's mouth sagged open, giving an unprepossessing view of streaky-bacon-flavour mulch. The Triv machine fell silent, and Teddy's hands twitched away from the buttons to his gun-handle. Allie enjoyed the moment, knowing everyone in the pub was knotted inside, wondering what the stranger—her friend, she realised—would do next. Gary Chilcot, a weaselly little Maskell hand, slipped away, into the back bar where the Squire usually drank.

"How d'ye do, Goodman," said Janet, stretching thin red lips around dazzling teeth in a fox smile. "What can I do you for?"

"Bells. And Tizer for Allie here."

"She'm underage."

"Maskell won't mind. We're old friends."

Janet fetched the whisky and the soft drink. Lytton looked at the exposed nape of her neck, where wisps of hair escaped, and caught the barmaid smiling in the mirror, eyes fixed on his even though he was standing behind her.

Lytton sipped his whisky, registering the sting in his eyes.

Janet went to the jukebox and put on Portishead. She walked back to the bar, almost dancing, hips in exaggerated motion. Music insinuated into the spaces between them all, blotting out their silent messages.

The door opened and Reeve Draper came in, out of breath. He had obviously been summoned.

"I've been meaning to call again on Goodwife Ames," he said to Lytton, not mentioning that when last he had seen Lytton the newcomer was on the ground with a bullethole in his shoulder put there by the Reeve's Constable. "Tony Jago, the Traveller Chieftain, has escaped from Glastonbury with a band of sheep-shaggin', drug-takin' gyppos. We'm expecting raids on farms. Susan should watch out for them. Bad lot, gyppos. No respect for property. They'm so stoned on dope they'm don't know what they'm doin'."

Lytton took a marijuana leaf badge from his pocket. One of the emblems pinned to the poncho left in the ravaged garden. He tossed it into Terry Gilpin's scrumpy.

"Oops, sorry," he said.

This time, Terry went for his gun and fumbled. Lytton kicked the stool from under him. Terry sprawled, choking on crisps, on the floor. With a boot-toe, Lytton pinned Terry's wrist. He nodded to Allie, and she took the gun away. Terry swore, brow dotted with cider-stinking sweat bullets.

Allie had held guns before, but not since Susan took her in. She had forgotten how heavy they were. The barrel drooped even though she held the gun two-handed, and accidentally happened to point at Terry's gut.

"If I made a complaint against this man, I don't suppose much would happen."

Draper said nothing. His face was as red as strawberry jam.

"I thought not."

Terry squirmed. Teddy gawped down at his brother.

Lytton took out his gun, pointed it at Teddy, said "pop," and put it back in its holster, all in one movement, between one heartbeat and the next. Teddy goggled, hand hovering inches away from his own gun.

"That was a fair fight," Lytton said. "Do you want to try it again?"

He let Terry go. Rubbing his reddened wrist, the Maskell man scurried away and stood up.

"If'n you gents got an argument, take it outside," Janet said. "I've got regulars who don't take to ruckus."

Lytton strolled across the room, towards the back bar. He pushed a door with frosted glass panels, and disclosed a small room with heavily-upholstered settees, horse-brasses on beams and faded hunt scenes on the wallpaper.

The Squire sat at a table with papers and maps spread out on it. A man Allie didn't know, who wore a collar and tie, sat with him. Erskine was there too, listening to Gary Chilcot, who had been talking since he left the bar.

The Squire was too annoyed to fake congeniality.

"We'd like privacy, if you please."

Lytton looked over the table. There was a large-scale survey map of the area, with red lines dotted across it. The corners were held down by ashtrays and empty glasses. The Squire had been illustrating some point by tapping the map, and his well-dressed guest was frozen in mid-nod.

Lytton, stepping back from the back bar, let the door swing closed in the face of Erskine, who was rushing out. A panel cracked and the Constable went down on his knees.

Allie felt excitement in her water.

Terry charged but Lytton stepped aside and lifted the Maskell man by the seat of his britches, heaving him up over the bar and barrelling him into the long mirror. Glass shattered.

Janet Speke, incandescent with proprietary fury, brought out a shotgun, which Lytton pinned to the bar with his arm.

"My apologies, Goodwife. He'll make up the damage."

There was nothing in the barmaid's pale blue eyes but hate. Impulsively, Lytton craned across and kissed her full on the lips. Hot angry spots appeared on her cheeks as he let her go. He detached her from the shotgun.

"You should be careful with these things," he said. "They're apt to discharge inconveniently if mishandled."

He fired both barrels at a framed photograph of Alder's victorious skittles team of '66. The noise was an astounding crash. Lytton broke the gun and dropped it. Erskine, nose bloody in his handkerchief, came out of the back bar with his Webley out and cocked.

This time, it was different. Lytton was armed.

Despite the hurt in his left shoulder, Lytton drew both his pistols in an instant and, at close range, shot off Erskine's ears. The Constable stood, appalled, blood pouring from fleshy nubs that would no longer hold his helmet up.

Erskine's shot went wild.

Lytton took cool aim and told the Constable to drop his Webley.

Erskine saw sense. The revolver clumped on the floor.

In an instant, Lytton holstered his pistols. The music came back, filling the quiet that followed the crashes and shots. Terry moaned in a heap behind the bar. Janet kicked him out. Erskine looked for his ears.

Lytton took another sip of Bells.

"Very fine," he commented.

Janet, lipstick smeared, touched her hair, deprived of her mirror, not knowing where free strands hung.

Lytton slipped a copper-coloured ten shilling note onto the bar.

"A round of drinks, I think," he said.

Danny Keough smiled and shook an empty glass.

✦

Outside, in the car park of The Valiant Soldier, Allie bubbled over. It was the most thrilling thing. To see Terry hit the mirror, Teddy staring at a draw he'd never beat, the Reeve helpless, Janet Speke and the Squire in impotent rage and, best of all, Barry Erskine with his helmet-brim on his nose and blood gushing onto his shoulders. For a moment, Alder was like *The Archers*, and the villains were seen off.

Lytton was sombre, cold, bravado gone.

"It was just a moment, Allie. An early fluke goal for our side. They still have the referee in their back pocket and fifteen extra players."

He looked around the car park.

"Any of these vehicles unfamiliar?"

Maskell's ostentatious Range Rover was parked by Janet's pink Vauxhall Mustang. The Morris pick-up was the Gilpins'. The Reeve's panda car was on the street. That left an Austin Maverick Allie had never seen before. She pointed it out.

"Company car," he said, tapping the windshield.

The front passenger seat was piled with glossy folders that had 'GREAT WESTERN RAILWAYS' embossed on their jackets.

"The clouds of mystery clear," he mused. "Do you have one of your nails?"

Puzzled, she took a nail from her purse and handed it over.

"Perfect," he said, crouching by the car door, working the nail into the lock. "This is a neat trick you shouldn't learn, Allie. There, my old sapper sergeant would be proud of me."

He got the door open, snatched one of the folders, and had the door shut again.

They left in a hurry, but slowed by the bus stop. The rusting shelter was fly-posted with car-boot sale announcements. Lytton sagged. His shirt-shoulder spotted where his wound had opened again. Still, he was better off than Earless Erskine.

"It's choo-choos, I'll be bound," he said. "The track they run on is always blooded."

There was activity at the pub as Maskell's party loped past the village oak into the car park. Maskell was in the centre, paying embarrassed attention to his guest, who presumably hadn't expected a bar brawl and an ear-shooting to go with his ploughman's lunch and a lecture on local geography.

The outsider got into his Maverick and Maskell waved him off. Then, he started shouting at his men. Allie smiled to hear him so angry, but Lytton looked grim.

That evening, after they had eaten, Lytton explained to Susan, showing her the maps and figures. Allie struggled to keep up.

"It's to do with Railway Privatisation," he said. "The measures that came in after the War, that centralised and nationalised so many industries, are being dismantled by the Tories. And private companies are stepping in. With many a kickback and inside deal."

"There's not been a railway near Alder for fifty years," Susan said.

"When British Rail is broken up, the companies that have bits of the old network will be set against each other like fighting dogs. They'll shut down some lines and open up others, not because they need to but to get one over on the next fellow. GWR, who are chummying up with the Squire, would like it if all trains from Wessex to London went through Bristol. They can up the fares, and cut off the Southeastern company. To do that, they need to put a branch line here, across the Southern edge of Maskell's farm, right through your orchard."

Susan understood, and was furious.

"I don't want a railway through my farm."

"But Maskell sees how much money he'd make. Not just from selling land at inflated prices. There'd be a watering halt. Maybe even a station."

"He can't do the deal without Gosmore Farm?"

"No."

"Well, he can whistle 'Lillibulero'."

"It may not be that easy."

The lights flickered and failed. The kitchen was lit only by the red glow of the wood fire.

"Allie, I told you to check the generator," Susan snapped.

Allie protested. She was careful about maintaining the generator. They'd once lost the refrigerator and had a week's milk quota spoil overnight.

Lytton signalled for quiet. He drew a gun from inside his waistcoat. Allie listened for sounds outside.

"Are the upstairs windows shuttered?" Lytton asked.

"I asked you not to bring those things indoors," Susan said, evenly. "I won't have guns in the house."

"You soon won't have a choice. There'll be unwelcome visitors."

Susan caught on and went quiet. Allie saw fearful shadows. There was a shot and the window over the basin exploded inwards. A fireball flew in and plopped onto the table, oily rags in flames. With determination, Susan took a flat breadboard and pressed out the fire.

Noise began. Loudspeakers were set up outside. Music hammered their ears. The Beatles' 'Helter Skelter.'

"Maskell's idea of hippie music," Lytton said.

In the din, gunshots spanged against stones, smashed through windows and shutters.

Lytton bundled Susan under the heavy kitchen table, and pushed Allie in after her.

"Stay here," he said, and was gone upstairs.

Allie tried putting her fingers in her ears and screwing her eyes shut. She was still in the middle of the attack.

"Is Maskell going to kill us?" she asked.

Susan was rigid. Allie hugged her.

There was a shot from upstairs. Lytton was returning fire.

"I'm going to help him," Allie said.

"No," shouted Susan, as Allie slipped out of her grasp. "Don't…"

She knew the house well enough to dart around in the dark without bumping into anything. Like Lytton, she headed upstairs.

From her bedroom window, which had already been shot out, she could see as far as the treeline. There was no moon. The Beatles still screamed. In the orchard, fires were set. Hooded figures danced between the trees, wearing ponchos and beads. She wasn't fooled. These weren't Jago's Travellers but Maskell's men.

Allie had to draw the line here. She and Susan had been pushed too far. They'd lost men to Maskell, they wouldn't lose land.

A man carrying a fireball dashed towards the house, aiming to throw it through a window. Allie drew a bead with her catapult and put a nail in his knee. She heard him shriek above the music. He tumbled over, fire thumping onto his chest and spreading to his poncho. He twisted, yelling like a stuck pig, and wrestled his way out of the burning hood.

It was Teddy Gilpin.

He scrambled back, limping and smouldering. She could have put another nail in his skull.

But didn't.

Lytton was in the hallway, switching between windows, using bullets to keep the attackers back. One lay still, face-down, on the lawn. Allie hoped it was Maskell.

She scrambled out of her window, clung to the drainpipe, and squeezed into shadows under the eaves. Like a bat, she hung, catapult dangling from her mouth. She monkeyed up onto the roof, and crawled behind the chimney.

If she kept them off the roof, they couldn't get close enough to fire the house. She didn't waste nails, but was ready to put a spike into the head of anyone who trespassed. But someone had thought of that first. She saw the ladder-top protruding over the far edge of the roof.

An arm went around her neck, and the catapult was twisted from her hand. She smelled his strong cider-and-shit stink.

"It be the little poacher," a voice cooed.

It was Stan Budge, Maskell's gamekeeper.

"Who'm trespassin' now?" she said, and fixed her teeth into his wrist.

Though she knew this was not a game, she was surprised when Budge punched her in the head, rattling her teeth, blurring her vision. She let him go. And he hit her again. She lost her footing, thumped against tiles and slid towards the gutter, slates loosening under her.

Budge grabbed her hair.

The hard yank on her scalp was hot agony. Budge pulled her away from the edge. She screamed.

"Wouldn't want nothing to happen to you," he said. "Not yet."

Budge forced her to go down the ladder, and a couple of men gripped her. She struggled, trying to kick shins.

Shots came from house and hillside.

"Take her round to the Squire," Budge ordered.

Allie was glad it was dark. No one could see the shamed tears on her cheeks. She felt so stupid. She had let Susan down. And Lytton.

Budge took off his hood and shook his head.

"No more bleddy fancy dress," he said.

She had to be dragged to where Maskell sat, smoking a cigar, in a deckchair between the loudspeakers.

"Allison, dear," he said. "Think, if it weren't for the Civil War, I'd *own* you. Then again, at this point in time, I might as well own you."

He shut off the cassette player.

Terry Gilpin and Barry Erskine—out of uniform, with white lumps of bandage on his head—held her between them. The Squire drew a long thin knife from his boot and let it catch the firelight.

Maskell plugged a karaoke microphone into the speaker.

"Susan," he said, booming. "You should come out now. We've driven off the gyppos. But we have someone you'll want to see."

He pointed the microphone at her and Terry wrenched her hair. Despite herself, she screamed.

"It's dear little Allison."

There was a muffled oath from inside.

"And your protector, Captain Lytton. He should come out too. Yes, we know a bit about him. Impressive war record, if hardly calculated to make him popular in these parts. Or anywhere."

Allie had no idea what that meant.

"Throw your gun out, if you would, Captain. We don't want any more accidents."

The back door opened, and firelight spilled out. A dark figure stepped onto the verandah.

"The gun, Lytton."

A gun was tossed down.

Erskine fairly slobbered with excitement. Allie felt him pressing close to her, writhing. Once he let her go, he would kill Lytton, she knew.

Lytton stood beside the door. Another figure joined him, shivering in a white shawl that was a streak in the dark.

"Ah, Susan," Maskell said, as if she had just arrived at his Christmas Feast. "Delighted you could join us."

Maskell's knifepoint played around Allie's throat, dimpling the skin, pricking tinily.

In a rush, it came to her that this had very little to do with railways and land and money. When it came down to it, the hurt Maskell fancied he was avenging was that he couldn't have Susan. Or Allie.

Knowing why didn't make things better.

Hand in hand, Lytton and Susan came across the lawn. Maskell's men gathered, jeering.

"Are you all right, Allison?" Susan asked.

"I'm sorry."

"It's not your fault, dear."

"I have papers with me," Maskell said, "if you'd care to sign. The terms are surprisingly generous, considering."

Lytton and Susan were close enough to see the knife.

"You sheep-shagging bastard," Susan said.

Lytton's other gun appeared from under her shawl. She raised her arm and fired. Allie felt wind as the bullet whistled past. Maskell's jaw came away in a gush of red-black. Susan shot him again, in the eye. He was thumped backwards, knife ripped away from Allie's throat, and laid on the grass, heels kicking.

"I said I didn't like guns," Susan announced. "I never said I couldn't use one."

Lytton took hold of Susan's shoulders and pulled her out of the way of the fusillade unleashed in their direction by Budge and Terry Gilpin.

Allie twisted in Erskine's grasp and rammed a bony knee between his legs. Erskine yelped, and she clawed his ear-bandages, ripping the wounds open.

The Constable staggered away, and was peppered by his comrades' fire. He took one in the lungs and knelt over the Squire, coughing up thick pink foam.

In a flash of gunfire, Allie saw Lytton sitting up, shielding Susan with his body, arm outstretched. He had picked up a pistol. The flashes stopped. Budge lay flat dead, and Gilpin gurgled, incapacitated by several wounds. Lytton was shot again too, in the leg.

He had fired his gun dry, and was reloading, taking rounds from his belt.

Car-lights froze the scene. The blood on the grass was deepest black. Faces were white as skulls. Lytton still carefully shoved new bullets into chambers. Susan struggled to sit up.

Reeve Draper got out of the panda car and assessed the situation. He stood over Maskell's body. The Squire's face was gone.

"Looks like you'm had a bad gyppo attack," he said.

Lytton snapped his revolver shut and held it loosely, not aiming. The Reeve turned away from him.

"But it be over now."

Erskine coughed himself quiet.

Allie wasn't sorry any of them were dead. If she was crying, it was for her father, for the chickens, for the vegetable garden.

"I assume Goodwife Ames no longer has to worry about her cows being destroyed?" Lytton asked.

The Reeve nodded, tightly.

"I thought so."

Draper ordered Gary Chilcot to gather the wounded and get them off Gosmore Farm.

"Take the rubbish too," Susan insisted, meaning the dead.

Chilcot, face painted with purple butterflies, was about to protest but Lytton still had the gun.

"Squire Maskell bain't givin' out no more pay packets, Gary," the Reeve reminded him.

Chilcot thought about it and ordered the able-bodied to clear the farm of corpses.

Allie woke up well after dawn. It was a glorious spring day. The blood on the grass had soaked in and was invisible. But there were windows that needed mending.

She went outside and saw Lytton and Susan by the generator. It was humming into life. Lytton had oil on his hands.

In the daylight, Susan seemed ghost-like.

Allie understood what it must be like. To kill a man. Even a man like Squire Maskell. It was as if Susan had killed a part of herself. Allie would have to be careful with Susan, try to coax her back.

"There," Lytton said. "Humming nicely."

"Thank you, Captain," said Susan.

Lytton's eyes narrowed minutely. Maskell had called him Captain.

"Thank you, Susan."

He touched her cheek.

"Thank you for everything."

Allie ran up and hugged Lytton. He held her too, not ferociously. She broke the embrace. Allie didn't want him to leave. But he would. The Norton was propped in the driveway, wheeled out beyond the open gate. He walked stiffly away from them and straddled the motorcycle. His leg wound was just a scratch. Allie and Susan followed him to the gate. Allie felt Susan's arm around her shoulders.

Lytton pulled on his gauntlets and curled his fingers around the handlebars. He didn't wince.

"You're Captain UI Lytton, aren't you?" Susan said.

There was a little hurt in his eyes. His frown-lines crinkled.

"You've heard of me."

"Most people have. Most people don't know how you could do what you did in the War."

"Sometimes you have a choice. Sometimes you don't."

Susan left Allie and slipped around the gate. She kissed Lytton. Not the way Lytton had kissed Janet Speke, like a slap, but slowly, awkwardly.

Allie was half-embarrassed, half-heartbroken.

"Thank you, Captain Lytton," Susan said. "There will always be a breakfast for you at Gosmore Farm."

"I never did give you the ten shillings," he smiled.

Allie was crying again and didn't know why. Susan let her fingers trail through Lytton's hair and across his shoulder. She stood back.

He pulled down his goggles, then kicked the Norton into life and drove off.

Allie scrambled through the gate and ran after him. She kept up with him, lungs protesting, until the village oak, then sank, exhausted, by the curb. Lytton turned on his saddle and waved, then was gone from her sight, headed out across the moors. She stayed, curled up under the oak, until she could no longer hear his engine.

THE WHITE STUFF

BY PETER F. HAMILTON
AND GRAHAM JOYCE

Nigel Finchley first blinked into the gleam on his way into the City, where he high-rolled other people's money on the trading floor. A Nimbus owner himself, he cast an appreciative gaze over the classic Lotus Esprit swishing up beside him at the traffic lights. Its engine purred with deliberate, sexual rhythm as the brunette Trust Fund Babe behind the wheel toed the accelerator in provocation.

But when he tried to eye-photo her silhouette, the glare coming off the Esprit's ice-blue paintwork defeated him. Squinting to filter out the reflected sunlight he realized just how mirror-bright that polish was. The Esprit had a sunbeam corona all of its own, making the rest of the queuing cars dull, mundane. Money, he told himself, money lays that kind of gleam on everything it touches.

The Esprit surged forward in a burst of arrogant power.

Nigel watched it go, thoughts contaminated by low-level resentment. Later he saw the gleam again. A Piccolo this time. Nothing wrong with Piccolos; MG versions were decent sporty runabouts. But they shouldn't gleam, not like that. He watched it pass. The Piccolo went gliding down the street with unnatural elegance.

His curiosity was roused. Almost unconsciously he began searching the traffic for more, and spotted another three examples before he swung down into the company's underground car park. Five extraordinarily gleaming cars out of a near-gridlocked city.

Nigel's regular lunchtime pub was The Swan, perched alongside a canal restored by a government benefit-earn scheme. Smartened up beyond the pocket of its original water-traffic clientele, serving a noveau cuisine menu, it had achieved a reputation equal to any of the area's contemporary wine bars, sucking in a whole strata of City financiers, the players of digital money. It had a whitewashed facade with a frieze of iron-rimmed cartwheels bolted onto the brick, and hanging baskets adorning the taproom door. The landlord served real ale from wooden kegs, and carrot juice from liquidisers with a sound like a dentist's drill. A large parking lot round the back was bordered by a high redbrick wall. It couldn't be seen from the street.

Nigel coasted the Nimbus into a spare slot, turned off the ignition, and looked up to see her. Maybe sixteen years old with freckles and a riot of vivid copper hair in tiny corkscrew curls. Her adolescent breasts bobbled like half tennis balls under a scoopneck T-shirt; her faded denim microskirt offered him a grand view of her long, suntanned legs. Bright noonday sun made her hair blaze, halo fashion.

She would be one of the kids from the sink estate on the edge of the Capitalcorp's redevelopment incentive zone—a hellhole of squatters, dealers, pimps, and exo-European illegals—all trying to make the same quid washing windscreens. Nigel felt a hot jab of envy. Though he had everything she didn't have, he envied her youth. He envied her street-sassy. He envied, very badly, the twentysomething black guy lurking possessively a few paces back, and who would undoubtedly be screwing her.

Lovely big emerald eyes glittered at him. "Hi there, captain, wanna have your wheels gleamed?"

"Huh?"

"Gleamed." A blink of flawless white teeth. She proffered a little square of metal. Sunlight skipped across its metallic purple paint, dazzling.

"Let me see that." Taking the metal square from her he tried to stroke its coloured surface, to understand the texture, but his fingers slipped about as if it were coated in warm ice.

"What is that?" he asked.

"Micro-friction layer, captain. We'll wipe your bodywork down, and spray it on." She shook a grey aerosol can in his face; no brand name. "Dirt and water can't get a grip, so your shine's permanent, and rust don't get a look in, see?"

He couldn't take his eyes off her. "How long does it last?"

"Always. It's micro-friction, right? Can't rub it off."

He ran through the dubious logic, his eyes wandering down to her legs again. "How much?"

"Twenty five."

It even seemed reasonable. "Count me in. Cheque or card?"

"Aww, come on!"

"If you want cash I'll have to find a hole in the wall."

"Fine. Have your pint and slot your card. We'll have your wheels sorted for when you get back." She stuck two fingers in her mouth and whistled. "Got a live one!"

Her proto-gangsta boyfriend stepped over to the car, attempting a customer-friendly smile. On that face, it was never going to work. His head had been shaved in a chessboard pattern, with each square of hair sprouting a single stubby dreadlock. The clothes were ultra-trendy; heavy biker boots crushed the tarmac.

Uneasy prejudices started cattle-prodding Nigel's defence mechanisms awake. Sure the guy was well-dressed, but the hostility was as blatant as Nigel's own disapproval.

They looked at each other, silently negotiating a demilitarized truce for the duration of the gleaming. The black guy clicked his fingers, and a posse of kids solidified around the Nimbus. Seven of them, ranging from sixteen years down to about ten: black kids with locks, white kids with bent-nail tattoos, Asian kids, all loaded with buckets and sponges and can- do. The young redhead was already sauntering off after her next victim.

Nigel paused on the pub's doorstep, the slow-turning cogs in his brain winching a frown onto his face. It had been a very slick operation, way beyond any usual street-rat earner. He turned to look back. The little shits were gleefully spraying his car white, great sweeps of fuck-the-rich graffiti sizzling eagerly out of the grey cans, an oily foam contaminating the grilles, the hubs, the windscreen. It looked like the Nimbus was getting ready for a shave. He was about to scream at them, but his shout never got past the first syllable. The white foam was gradually turning translucent, smoothing out to form a thin, uniform coat, already delivering the gleam.

The redhead caught his eye, giving him a laughing thumbs up. Feeling hopelessly old and dumb (memory image kicking in: his father holding his first-ever CD up to the light in utter perplexity) he smiled back weakly and retreated into the pub. Pity there had been no brand name on the spray cans; the company who owned that process would be worth a dabble. Interesting.

Rewarded by that not-totally-innocent smile of hers, Nigel had promised the redhead to put the word round the trading floor. In two days all his smart colleagues were driving round in gleaming cars.

The Swan's landlord didn't object. Customers parked their cars and checked in for a drink while the kids sprang to work with sponges and spray cans. They had a regular production line going out back. Only once did the financier in him assert itself. How much should a micro-friction coating actually cost? Were the kids at The Swan pulling a fast one?

Nigel tried to price the coating at his Nimbus dealer's showroom. "What's micro-friction?" was the reply.

There were at least a dozen kids in The Swan's parking lot when he pulled in the following Monday. Five cars had their bonnets up, two kids to a car, doubled up over the radiator grilles, looking like they were being swallowed whole.

The redhead bounded over. Today she wore hip-hugging navy-blue shorts and a sleeveless white blouse, top four buttons undone. Boyfriend nowhere to be seen.

"Business is looking good," Nigel commented.

"I give people what they want."

He glanced up. He wondered exactly how old she was.

"Take you. I mean, these wheels of yours: seriously loaded." Her hand stroked the bonnet, coyly. "But you can't jack up the throttle cos of these speed laws. They ain't tailored to modern cars. They're antiques, thirties fodder."

"How do I just know you're going to sell me something else?"

Her answering grin was evil, moist tongue tip peeking out from the corner of her mouth. "Cos I like ringing your bell. I've got a zapper scrambler here that's got your name on it. You game?"

He tried to keep his eyes off the open buttons of her blouse. "Maybe. Speak to me."

"It screws the law's radar guns something chronic. First, a warning bleep from the laser detector at half a mile, then the LCD counts you down to ground zero. And the big plus: even if they do blast you, their read-out swears blind you're only doing a poxy twenty eight."

No more fines, no more penalty points. Every motorist's dream gadget. "How much?"

"Fifty."

He sighed. "I'm all yours."

The next day he drove into the tarmac wasteland of his local hypermarket's customer park and shot into a space near a trolley rack, tires crunching the litter of polystyrene wrappers.

"Wow, this is one totalled-out machine." The Nimbus's admirer was another girl: mid-teens, golden hair, dirty fingernails and white jeans as tight as a tourniquet. "I just bet you could go supersonic if it wasn't for those dumb speed limits."

He pointed to the newly installed LCD radar-trap warning on his dashboard. The girl shrugged and moved on. Looking out across the hypermarket park he could see nearly all of the cars sported microfriction coatings. Several cars in the row behind him had their bonnets raised, with kiddie teams slamming in zapper scramblers as though they were on a triple bonus productivity scheme. They also had a runner. A twelve-year-old boy collected the cash from the older kids, then disappeared into a graffiti-splashed alley at the rear of the park. A minute later he would reemerge with boxes of scramblers.

Nigel strolled over to the hypermarket entrance. Simon was sitting in his usual place beside the wire baskets, wrapped in a thick Oxfam duffle coat despite the warm sun. Scuffed wraparound sunglasses made him look like a washed up Terminator. He was playing his flute, a tired golden Labrador guarding a threadbare cap with a few coins in.

"Morning, Simon."

"That you, Mr Finchley?" Simon asked.

"Indeed it is." Nigel found a coin and bent down as if to drop it into the cap. He made the coin chink quietly, to disguise the fact that he kept it in his hand. It was a nothing- for-nothing world, in which Nigel was prepared to donate to the blind beggar no more than the sound of his money.

"Thank you, sir."

"My pleasure."

As Nigel stood up he saw the runner on the other side of the road. He was sure the boy had looked away quickly, a subliminal impression of a guilty start.

When Nigel had negotiated the maze of dingy backstreets at the side of the hypermarket he found the other end of the alley was blocked by a hired Transit van. A young black man was sitting in the driver's seat, flicking through a tabloid newspaper.

Nigel ambled past, snatching a glimpse of the runner returning to the car park, a scrambler box tucked under each arm. Another young man stood beside the van's rear doors.

The pair of them could have been cousins of the redhead's boyfriend. There was something shared between them; it wasn't so much a physical characteristic, more an attitude. Not arrogance exactly. Confidence. They possessed confidence.

Unenlightened, Nigel moved on. If they'd got the whole day's cash taking in the van, they'd be nervous about people who loitered.

Wednesday saw the redhead in a black one-piece cycling uniform. Nigel couldn't understand how she'd got into it, the fabric was already stretched to its limit.

Her knowing grin was becoming a little too familiar.

"Have you ever heard of the term 'market saturation'?" he asked before she made her pitch.

She stuck her tongue out, awesomely childlike. "Nah. Have you ever heard of an encryption-buster?" Resting in her palm was a matte-black box the size of a blockbuster paperback. There was a small keyboard on top. "Unkink everything the satellites beam down: Disney through to Movie Channel, Hot Dutch, the works, no smart card required. You'll never cop for a subscription charge again."

"Very nice. Where did you get it?"

"Bloke in a pub."

Acquisitive lust began to gnaw. The cost of a decoder card for his Globecast system was criminal. "How much?" It seemed to be the one consistent phrase he spoke to her.

"Fifty for cash."

"Sold. Come and have a drink with me."

She looked over her shoulder and thought for a moment. "Sure."

Nigel lived alone in his Docklands condominium. There were plenty of trees lining the empty streets, and no delinquents since the entire area was security-ringed and patrolled. Sometimes the only movement in the whole neighbourhood would be that of a scrap of newspaper blowing down the achingly new concrete walkways.

He told himself he still enjoyed the single life. He had enough friends in the same situation to make a cosy self- reinforcing group when they spent their weekend nights on the town. Jannice had been his last permanent attachment, though the relationship had broken down on a dispute over language.

"Please don't refer to me as your 'partner' or your 'girlfriend.' It's insulting and demeaning."

"So what are you, then?"

"Nothing which implies a contract."

I'm being outmoded before I'm thirty, Nigel had thought at the time. Jannice was four months ago. There had been other girls since, one night stands picked up while clubbing, a friend's younger sister. But his job on the trading floor was secure, which in itself was a bonus these days. The City and the new government were still eying each other wearily across the political divide; but apart from a little ideologically symbolic blood-letting among the fat cats of the utilities in the first six months after the election, there had been no incursions, no major campaigns led by reforming chancellors. The sheer voltage of money flowing through the cables of the City's finance web was so great nobody was going to risk shorting it out. So Nigel and his kind were still allowed to play their fast, adrenaline-high game.

Bathing in the timid blue phosphorescence of the monitors, he drank down information, hungry for the elusive patterns that bespoke success. When he found one, a bond, a rights issue, a commodity, he pumped money into the precious new find, guarding the knowledge until the stock rose and his investment grew ripe for harvest. He bred money from money, a nexus between data and currency arranging diabolical matings. Always on the hunt for new brides. A search he could run on autopilot these days. Same as his life.

And so unlike Miranda, the young redhead who had unsettled him. A teeny-rebel, making money and having fun, delighting in life. She made him realize that his own secret promises to himself had been broken; that his technicolour dreams had been pawned to pay for a permanent place on the trading floor. Freshness for stability.

That drink in The Swan had turned into two before she would even tell him her name. Then when he'd offered to take her to dinner she'd narrowed her eyes at him.

"It's a good thing I'm older than I look," she said, fingering the stem of her glass.

"Why? How old do you look?" The stupidity of this question didn't strike him until long afterwards.

He'd collected her from The Swan later that evening. Later than he intended, actually. The floor had gone through one of those unexplained jittery days; as if nervousness had suddenly mutated into an airborne virus, circulated by the slow-spinning rooftop fans of the City's air conditioners. End-of-month figures showed African imports of electronics were down, reducing the continent's borrowing. Rumour-quakes ran gleefully through the money market. Several blue chip companies turned slightly pale. He hated days like that, hated the disorder.

Miranda had waited, though, an encouraging measure of her eagerness to sample the good life. She'd applied too much make-up, and that a little carelessly, but it didn't diminish her. He broke the speed limit thanks to his new box and tried to impress her by taking her to a Chelsea restaurant supposedly used by Princess Di, knowing she'd be completely out of her depth. Princess Di wasn't in, but it looked as if the maitre d' was operating a beauty code for patrons.

"Hot dump this, eh?" Miranda said as they sat. Her gaze hardened as she took in the designer dresses by Lang, Versolato, Rocha, and Westwood. Her own dress was some not-quite-Goth purple velvet with a low front and black lace sleeves.

"All the best TFBs come here," Nigel assured her.

"?"

"Trust Fund Babes. Never done a day's work in their lives."

"You normally go out with women like that?"

"Only when they're slumming. They tend to go for farmers who own half of Sussex."

Miranda ordered the same dishes as him; she watched him carefully when the food arrived, mirroring his movements, and choosing the same cutlery. It wasn't as amusing as he'd expected. She was so bloody determined. He knew that for the rest of her life now she would always select the right fork, would tilt her soup bowl away from her.

"Another bottle of champagne," Nigel said to the wine waiter.

"Don't waste your folding," Miranda said. "I've already decided to fuck you."

They lay on his bed, the curtains of his room drawn back and the strange spectral light from the flashing and steaming Cesar Pelli tower reflecting on the perspiration of their naked bodies. He tried to get her to tell him where she got all the strange new merchandise. One slip of the tongue, one name, was all he needed to make his killing on the floor.

"Secret."

"Ah, come on!" His voice mellowed out. "Between us?"

She flinched, confused and vulnerable. "Don't know much. Honest. All I know, it's called *afto-aspro*. Same stuff in the electronics as does for the gleam."

He puzzled over that. "*Afto-aspro?*"

"Yeah. Ilkia says that's what the Greeks called it. Means white stuff."

"It's manufactured in Greece?"

"No. That's just where it turned up first. Couple of weeks back."

"So where does it come from originally?"

"Ilkia says the exos brought it in with them when they come over from Africa. It's all over the estate now."

"Who's this Ilkia character?"

"Mate of mine. He'd kill me if he knew I was here."

"Jealous type, is he?"

"Not that…. Well, sometimes. He gets us the *afto-aspro* gear, see. I have to keep him sweet; and he don't like the likes of you."

"White?"

"Nah. Rich. Ilkia says companies like yours are the generals on your side of the class war."

"Oh, Jesus wept. Look, does this Ilkia know who's manufacturing *afto-aspro*?"

"Dunno. He never says much about it, just bangs on about how it's gonna make things different for us. All the global capitalist state is gonna get whacked. It'll start with the electronics companies, and when they go, they'll bring everything else down with them."

"I think your friend is talking out of his arse. He really doesn't know much about enterprise economics. The electronics industry is a perpetual war of innovations and next generation chips. That's what makes the companies so dynamic, and strong. One new gizmo isn't going to bring civilization to a halt."

"We don't want to halt it, just change it."

"So, broadly speaking, would you describe yourself as an anarchist, or just another rainbow Nazi?"

"Don't say stuff like that. Nigel, be straight, d'you think it's possible for someone like you to love someone like me?"

He grinned savagely, she didn't get it—too young. "Let me show you instead."

When she was finally asleep he went through her bag. Usual teenage junk, except for a wad of seven hundred quid in new twenties held together by elastic bands. No hint to the origin of *afto-aspro*.

The only thing he didn't understand was a slim oblong of plastic with chrome-silver surfaces, about the same size as a credit card.

"I want a favour from you," he asked her over breakfast.

Miranda giggled. "I thought I did all that last night."

"I want you to find out more about *afto-aspro*."

"You got all I know."

"Listen, you want to be like me, to run in my world, move in my circles?"

"Maybe." She nodded, face all dumb and serious. "You ain't how Ilkia said you would be. And this place… I had a good time last night, Nigel. Honest. That's not greedy, is it? Not to want that?"

"Nothing like. Motivation makes the world go round. You have to give people incentives. As a race we need to create and achieve; the alternative is stagnation."

"Right. Yeah."

"This is your chance to achieve, Miranda. You can come in with me, I can make you part of my deal. All I need is the name of the company which produces *afto-aspro*. I can buy up their stock and cut myself in for a big percentage. It'll be like knowing the lottery roll-over numbers in advance. Now do you want a piece of that? Do you want last night to be every night?"

"You shitting me?"

"Just bring me the information."

Trouble mugged Nigel as soon as he reached the floor. Everyone was on their feet screaming into telephones. The markets were going crazy. High Street banks had reported a massive surge in demand for gold sovereigns. There was no logical reason for it. Of course the banks didn't stock the coins, they had to be ordered from the Bank of England.

There was a similar demand sweeping the entire European mainland. Not from dealers, but from the public. Gold prices grew by the minute. Nobody knew what was happening. Yesterday's nervousness blossomed out into full-scale panic.

Six hours later everyone held their breath as New York started trading. Wall Street dived straight into the gold market. And Nigel found out the true meaning of pandemonium.

After a terrible day he washed up at The Swan, hoping to find Miranda there. She wasn't. But a fourteen-year-old girl wanted to sell him an emax.

"A what?"

Freckles crinkled against spots as she smiled. "An energy matrix, what they used to call a battery." She showed him a small fat cylinder: black, glossy, seamless. "The outside casing is a solar collector, see? Ninety-five per cent conversion efficiency. You just have to leave it in the sun and it'll recharge in a couple of hours."

Ten quid each. He bought six to power his ghetto blaster.

Next morning the public's thirst for gold had increased. The Chancellor appeared on the lunchtime news to try to calm people, assuring them that the Bank of England had enough reserves to cope with the unexpected consumer-led boom, and no restrictions were even being contemplated. The interviewer's questions about the economy starting to downturn in such a climate were brusquely dismissed as scaremongering.

Nigel couldn't concentrate that afternoon, despite all the floor supervisor's screams and threats to level their investments. He spent the time accessing share prices for electronics companies. What he found was more unsettling than any of the five-million-quid skeletons he had rattling round his accounts. Miranda had been right: the prices were slowly starting to drop. Worse, it was a global picture. Other analysts would be plotting the trend, the whole electronics section of the market would crash. If he just knew which name made *afto-aspro* he could pump millions into their stock and ride the storm's lightning.

Before he left work that evening a rumour swept the floor that cashpoint machines all over town had malfunctioned, dishing out three thousand pound windfalls to hundreds of lucky punters. The banks were closing down their hole-in-the-wall outlets until the electronics could be checked.

He started the drive home. The traffic all around him shone like a river of prismatic sunlight. Everyone, these days, gleamed.

Halfway to Docklands he saw three ten-year-old girls standing beside a building society's cashpoint. One of them had a silver card just like the one in Miranda's bag, which she shoved into the slot. Money started gushing out. The girls squealed excitedly, scooping it up.

Nigel parked and walked, soaking up the new fizz loose on the block. A knot of five boys loitered ahead of him. He had no doubt that the one with his back to him was an *afto-aspro* peddler. It was the clothes—bright, new, expensive. There was a glint of gold necklace chains exchanged for a slimline afto- aspro box. The peddler shook hands and departed; not getting three paces before more people buttonholed him.

Each of the boys he left behind registered an awed, vaguely guilty expression as they stared at their new *afto-aspro* box. Nigel followed them without a qualm.

It was crazy. They marched into a jeans shop that was chromed, strobed, and shaking with some classic Pulp. A bored girl assistant read a tatty X Libris paperback behind the counter. The boys clustered at the back, sunburst xenon pulses segmenting their movements to robotic jerks as they stuffed jeans into plastic carrier bags. Then they sauntered out casually. The assistant never even glanced up.

Nigel started after them, slackjawed; then he saw the two white pillars on either side of the open door. *Why not?* he thought. *If they can cobble up something that'll scramble a radar trap, then why not something to jam a shop's security tags?*

Outside on the pavement a man was struggling past, carrying what looked like a huge painting, an oblong of brown wrapping paper five foot by three foot, barely an inch thick. A furtive look in his eye suggested he'd just snapped up a bargain.

Nigel raised a finger. "Excuse me."

It was a flatscreen television. No need to use up valuable living room space with a bulky black cabinet, just screw it on the wall, neat and out of the way. The screen was edged with a solar collector frame, so it didn't need to be plugged into the mains. Better yet, the man confided, it wouldn't register on TV detector van equipment, no need for a licence. And all for a hundred pounds.

Nigel knew that the corporate giants like Sony, JVC, Goldstar, IBM, and Racal had spent most of the last decade and hundreds of millions of dollars into cracking the concept of flat wide screens.

The blue chips would be haemorrhaging white tonight.

"Mobile phone, mister?" A dreadlocked Asian boy smiled winningly; two missing front teeth turned him into a juvenile vampire. "It's got a floating clone number, you never get a bill. Twenty quid folding, or a sovereign."

"Sod off."

✦

Miranda was waiting for him on the leather settee in his lounge, denim shirt unbuttoned to show off a small black bra.

"How the fuck did you get in here?"

She grinned, and held up a snow-white version of his HiSecure infrared key. "Fair's fair, Nigel. You got the key to my panties."

"You shouldn't be carrying that kind of gadget round with you right now. People are starting to realize what *afto-aspro* can be made to do."

"Sure they are. They've seen what's coming; they're gonna be carrying themselves pretty soon. Just like Ilkia said. Have you seen the way it's taking off? Stuff's flooding out of the sink estates. There's kids in every city got a breeder chip now. Nobody's going to stop it."

He glanced up from the Twenties glitz-mirror cocktail bar. "A breeder chip? You found out something?"

For once the youthful confidence was missing. She shivered a little as she took the Pimms he'd mixed. "Yeah. I got Ilkia to take me to the Cameroon boys' squat. They showed me. All you need to make *afto-aspro* is a little chip of breeder and the right chemical junk for it to scoff. Theirs is hooked up to an old PC running a composition program. Do you see? You tell the *afto-aspro* what you want it to be, and it just fucking does it." She paused. "The function is hardformatted into the molecular structure. It can be anything you want."

Nigel sank deep into the leather settee. "Holy shit." No name. She'd told him there was no name, no single source, no stock to invest in. It was the end of the world.

She laughed, kitten spry again. "Isn't it beautiful? It was crude gear at first, like the gleam and the encryption-buster; but there are composition program upgrades coming in every hour now, downloaded through the Internet. People like the Cameroon boys are matching anything the big companies can build, and then some." Miranda threw her arms round him and tried to kiss him.

"You don't understand what you're doing, do you?"

"Yeah. I'm having a good time. I'm winning. Like you."

"Jesus Christ."

"So is that enough to make our deal? Are we hot now, or what?"

"What deal? You and your friends are screwing the world to death. Do you understand that? *To death!*"

She looked at him strangely, as if he'd missed some terribly obvious point. "Sure it's gonna be different for you. Your world's gonna be the same as mine, now. Didn't you realize that? That's why you wanted the deal, ain't it, so we could make a stash and clean out first?"

"Stash of *what*, you stupid bitch? Your bastards from Cameroon are wiping out the economy."

"No we ain't. We're just spreading it around a bit. Stopping things costing so much. *Afto-aspro* lets people like me have what you've got. No hurt in that. We'll all be better off."

"You understand nothing."

Miranda laughed and climbed onto his lap. "Kiss me Nigel. It's going to be a brand new day. A whole new world! We're gonna change everything! Ilkia says it's democratic electronics. He says that's what it was designed for, to give the world's poor what the rich Westerners enjoy. *Afto-aspro* don't kill, it helps everyone. Ilkia says it's only the old capitalist structure which is dying. There's still going to be an economy, but we're going to free it from banks and billionaires. Ilkia says- "

"Fuck what Ilkia says." Nigel surged upwards, violently pushing her away. She fell back heavily, cracking her head against the mahogany cocktail bar.

For the first time she seemed to realize his position. She looked confused. "I thought you and me had a thing."

"Don't be ridiculous."

Miranda went from confusion to plain hurt. She stroked strands of hair from her eyes, which were moist. "Here's a tip, Nigel," she said tearfully, going to the door. "Hang on to your gold watch, and your gold pen, and your tie-clip, and your gold neck chain. You'll need them. Credit's gonna get busted too, and that's what your kind live on. Cos we never get given any, do we?" Then she was gone.

He stood there for maybe half a minute. "Hell." He ran out into the landing, but the lift door was already closing. "Good luck, Miranda," he shouted after her. "I want to wish you good luck."

The TV news was overloading on *afto-aspro*. Half of the nine o'clock news was devoted to the new manufacturing revolution, hot young reporters tackling happy punters. Studio talking heads followed with hyper-cautious interpretations of its possible consequences. The end of large-scale industrialisation, the start of a true global village economy, green, clean, and noncompetitive.

"Bollocks," Nigel snapped. He fired the remote, wiping out the report. De-industrialisation wasn't the outcome, he was sure of it. Instinct was strong here, a lifetime of feeling the patterns spoke to him.

He pulled his laptop over, flipped it open, and after a while, began to type.

"Dump every electronics share we have," Nigel told Austin St. Clair next morning.

"Are you insane?" the supervisor bawled. Outside the glass walls chairs had become as redundant as silicon; everyone stood, shouting and waving at each other like bookies on speed. Gold was still rising. Governments were issuing optimistic forecasts—nobody was paying attention. VDUs were flashing up red starburst symbols as the electronics market crashed. "If we sell now, we'll lose over a quarter of a billion. Totally outside my authority. I won't be working the floor, I'll be fucking sweeping it!"

"We'll lose a lot more if we hang on to them."

"It'll bottom out. It looks bad, but it always bottoms out. Intel and Fujitsu have already announced they're going to start *afto-aspro* production; the rest will follow."

"It doesn't matter. Those kind of companies are obsolete now. It took an investment of half a billion dollars to build every new chip plant; and two years later the next generation processor would come along, and you'd have to rebuild. That's why only the rich countries ever produced the damn things. The whole point of *afto-aspro* is that it doesn't need that kind of investment. Right now, the dumbest people in the country are making *afto-aspro* in rooms full of cockroaches. Don't you get it? There is no more electronics industry. It just went the same way as gas lamps and vinyl records."

"Christ!" Austin thumped a fist on his desk, then jammed the grazed knuckles into his mouth. "Oh Christ, Nigel."

"It doesn't matter. We can get the company out. Buy engineering, the heavier the better. Firms which make big, solid, bulky metal products: ships, cranes, combine harvesters, bridges, cement lorries, steam rollers, trains, hell even washing machines are mostly mechanical. *Afto-aspro* can't replace that. And those shares are a good buy right now. Everyone was so keen to get into sunrise technologies and multimedia bollocks we ignored the fundamentals. If we move quick we can recoup our losses on electronics. But it's got to be now, Austin. The market will figure it out; capital is going to flood into that section. If you're smart, we'll be ahead of them."

Five months later his prediction had almost come true. There were no more electronics companies. There weren't any oil companies left either, thanks to the ubiquitous emax.

The *afto-aspro* spring had edged out the winter of faded technologies right across the UK. Vigorous new growth supplanted the obsolete structures and systems of yesterday. Solar collector panels were spread over roofs, replacing slates and tiles and thatch. Cars were fitted with electrolyte regenerator cells, kits which turned exhaust fumes back into petrol. That was just a stop-gap. Factories were already busy installing new production line facilities for vehicles which would be powered solely from an emax, their bodywork reverting to the Henry Ford bon mot of gleaming solar-collector black.

Afto-aspro was at the heart of it all, but the actual change, the physical adaptations, required manual labour, skilled and semi-skilled. Opportunity for all. People lost jobs, people found new ones. Unemployment only rose a couple of per cent. Nigel was laid off and re-employed on a freelance basis, lower income, inferior terms, poorer conditions. But at least he was still in work.

It was the day the gas network was due to be turned off permanently when Nigel glided his not-so-new Nimbus into the hypermarket park. He turned off the ignition and the perfect tone of the Sonic Energy Authority ebbed away; the *afto-aspro* MB (memory block) player had replaced CDs and cassettes. MBs had also replaced videos, games cartridges, and floppy disks; each cigarette-sized cylinder stored hours of data.

There were no kids lurking about ready to thrust the latest *afto-aspro* application into his face. He missed them somehow. But for fitting regenerator cells on cars you went to a garage; to wire your home up to a domestic emax a professional electrician was called in; any household gadget was grown to order in your local electrical store.

Blind Simon was hunched in his usual place beside the hypermarket entrance, coat buttoned up against the sweltering September sun, flute trilling gently. Nigel patted his pockets for a coin. There was a wad of notes in his wallet; for a couple of weeks transactions had been all in sovereigns, or jewellery, or even art; but with things settling down again people were accepting the promise of the Chief Cashier once more. Even cashpoints were coming back in use as banks replaced their old electronics with blocks of *afto-aspro*.

"Morning, Simon."

Simon smiled softly. He raised the scuffed old shades and looked straight at Nigel. "I always wondered what you looked like. I always wondered about the face of a man who would pull a shitty stunt like that, week in, week out."

The golden Labrador barked angrily.

Nigel stumbled a pace backwards, shock draining the heat from his blood. Simon's eye sockets were filled with balls of *afto-aspro*. No irises, no pupils, just blank white spheres. "Clever, isn't it?" the old tramp said. "The latest compositional program upgrade can design organic substitutes. Eyes are easy; all an eye does is convert photons to nerve impulses. Molecular filters like kidneys are a little more complicated, but they'll get there, I'm sure. After all, the only real work left these days is thinking."

"Christ almighty."

"How does your money sound these days, Mr Finchley?"

The heat returned to Nigel's blood as fast as it had left, burning his cheeks and ears. He almost sprinted back to the Nimbus.

✦

The trading floor was quiet. Half of the terminals were veiled below dust covers; the level of activity on the market no longer justified a full team of dealers. Those still on the company's payroll were a subdued, sober bunch intent on steering a steady course. The days of screaming out deals while holding three telephone handsets and reading five displays at once were long gone. After the holocaust of corporate casualties *afto-aspro* had inflicted on the global economy, what remained was rock solid stable. The international financial playing field still wasn't exactly level, but the disparity between the developed nations and what had been the Third World was a lot smaller. In fact, the distinction between the two was now measured in the amount of infrastructure a country had: industrial output per capita was approaching equilibrium. As people suddenly realized, the Third World had a lot of heavy engineering plants, most of them built by the multinationals who wished to exploit the cheap labour costs such countries boasted as their principal asset.

Nigel walked down the row of silent, blank computers knowing how grave-robbers must feel. Dealers just picked over the bones these days; they didn't control or dictate like before. But like everyone else, he'd adjusted. He was good at picking over bones, spotting the scraps of flesh. His position was almost as safe in the new order as it had been in the old.

"Some weird delegation in," one of the dealers muttered uneasily as Nigel sat at his station. "Austin's been talking to them for forty minutes."

Four people were sitting in Austin St. Clair's glass wall office. Not the usual collection of Armani and Yamamoto power suits either. Three black men and a red-headed white woman, all dressed in army surplus fatigues and combat boots.

Austin St. Clair caught sight of Nigel, and met him at the door. "Trouble," he announced bleakly before Nigel could get into the office. "Anarchist freaks from the London-Cameroon software collective. They've made an approach to our board for improving the trading floor's performance. So they say."

As the door closed behind Nigel, he glanced over at the intruders, and froze. It was Miranda. Miranda looking poised, taller, broader, delectable.

"Hi there, Nigel. It's been a while."

"Miranda?" He tried to retain his composure. "What's new?"

"Meet the boys from the collective. We're about to buy you out."

"Buy us out? Well, I see you joined the human race. I always said ex- anarchists make the best capitalists." He tried an uneasy laugh.

"Not quite." She glanced at Austin St. Clair as if to decide his trustworthiness, then shrugged and relaxed. "Actually, we're going to stamp you out, Nigel."

"No way."

"You like the way I look now?"

"What? Yes. Yes, I suppose so."

"It's the latest *afto-aspro*; our collective specialises in compositional programs for organic substitutes. I had a few implants."

"They suit you. You suit you."

"We don't really give a fuck about cosmetics, of course; but it helps screw up the income for private clinics. What we're concentrating on is providing enhanced automated intellectual services."

"What?"

"Neural networks. We grow *afto-aspro* brains, Nigel."

"The London-Cameroon collective has persuaded the board that their new *afto-aspro* development can handle the trading floor by itself," Austin St. Clair said grimly. "They're going to wire neural networks into our finance net, and replace the dealers."

"*What?*" Nigel yelped.

All four of the collective were smiling at him.

"We're using capitalism's own strength to break it, Nigel," Miranda said. "Capitalism fosters the culture of competition and achievement; so someone told me once. And in order to compete and achieve you've got to have the best product. Once we've installed the system here, the other financial companies will see how good it is, and they'll buy the same system for themselves. They'll be refined and polished and debugged until they can't be improved any further. Then everybody will have identical systems battling for the same business. It'll be the final stalemate; nobody will be able to win. You'll be levelled, Nigel. What you are now will cease to exist. It'll allow a social market to grow without interference."

"Banks and billionaires," he whispered.

"You got it, comrade. But before we sound the last crash, the collective would like to offer you a week's contract. You always said you were the best, Nigel, now here's your chance to prove it. Our neural networks need to learn the ropes, so who better than you to teach them their core program? They'll spend a week observing you deal, then take over. Our terms for your thought routines will be generous."

"You want my thought routines?"

"Yes. That's all you have left to sell, Nigel. The last aspiration."

"But what about me?" he yelled. "What about after?"

"Try earning a living," Miranda said. "I wish you luck. Really, a lot of luck."

A NIGHT ON THE TOWN

BY NOEL K. HANNAN

"El capitalismo convirta a Caracas en un inferno"
— graffiti on Caracas bus station

Miguel is trying so hard to impress her, he really is. He has greased his hair and brushed his teeth—twice, with the new American toothpaste that nanotechnically scours your mouth—and lightly rouged his cheeks. He is wearing his older brother's favourite outfit (Carlos would kill him if he came back from his school outward bound holiday on Margarita Island and found him wearing it)—nylon and leather parachutist's boots, baggy cotton pants and skinny-rib black T-shirt showing off his concave stomach and multicoloured Inca sunburst tattoo encircling his navel. He looks gorgeous, like a rich seventeen-year-old alone with a beautiful young woman in his family apartment in Nuevo Caracas should look. And *still* she is not impressed.

She sits in the moisture-slicked bay window, looking out over the firefly city as the sun is eaten by a storm sky, toying with a narcotic All-Day Sucker, her long brown legs dangling naked from the dramatic split in her halter-necked blood-red ball gown. She does not even flinch as the slam of thunder rocks the city. Maria is eighteen years old and a raven-haired Latin beauty. A year older than Miguel— it may as well be a hundred. She has made an art form of cynicism and world-weariness. The narcotic lollipop that Miguel bought her from a street vendor on their way here should be making her buzz. Instead, it appears to intensify her boredom.

Miguel is desperate. Maria is a goddess, her body curved and voluptuous. He very much wants to return to school on Monday and boast of his sexual adventures—which he will of course, even if he does not bed this impressively unimpressible siren. But the conditions are so right! His botanist parents away on a field trip in the rainforest—no school until Monday morning—a Saturday night city stretching and limbering thirty stories beneath them—his creditocard full and active (*praise Jesus!*) and his father's brand-new red Ford Machos "Matador" Special Edition waiting in the basement garage. They can go anywhere and do anything. God, what will it take to make this woman horny?

He slumps in the formocouch and watches her. She slips from the window sill with a bored sigh and is momentarily highlighted by sheet lightning as the storm breaks over Nuevo Caracas, wild photons dopplering her bare shoulders with jungle tiger patterns. She moves toward him with liquid grace, bare feet padding on thick carpet. She kneels at his feet and places her hands firmly on his splayed thighs. He stiffens.

"I need to eat," she breathes, running her tongue across her glossy lips. A faint whiff of lemon drifts from her breath, the scent of the narcotic.

"Take me to dinner," she insists, settling back on her haunches like a karate fighter awaiting a bout.

He swallows hard before answering her.

"What would you like to eat, Maria?"

Her dark eyes flare. The first sign of passion he has seen since he brought her here.

"Something special," she purrs. "Something unusual. Something exotic."

As she speaks her fingers trace the inside of his thighs. He feels the pressure of her sharp nails through the thin cotton pants.

"Take me somewhere different, Miguel." It is, he thinks, the first time she has spoken his name. She makes it sound like treacle being poured on velvet. *Miguel. Miguel. Miguel.*

So, he thinks, let's recap. Saturday night. Parents in the forest. Carlos on Margarita. Apartment free. City buzzing. Ford Machos "Matador" Special Edition in garage, keycard in pocket. Money no problem. Beautiful, high-as-a-kite Maria Del Fuego in a thigh-split red cocktail dress on her knees—*on her knees!*—in front of him. There is, of course, as mad as it seems, only one serious course of action.

There are myths and legends that permeate Nuevo Caracas like no other city on Earth. In this place where the rainforest hugs the cyberscraper as it smothers the congested, disease-ridden barrio, the *brujo* or witch-doctor of the forest tribes is as respected as the Catholic priests who ply their trade from streetside booths, whispering Latin mantras from under smog masks and rain capes. There is a story that Miguel has heard many times and which he is frantically trying to recall the details of now. The story concerns a *brujo* in the southern part of the barrio that rings the cybercity. The *brujo* owns a restaurant situated in the abandoned ruin of a nineteenth-century mission church, a tiny collection of crates and candles stuck in wax-encrusted wine bottles, huddled beneath corrugated plastic sheeting. In this "restaurant" the *brujo* weaves culinary magic that brings the affluent down from their crystal towers to run the gauntlet of muggers and lepers and beggars, of car thieves and body-part bootleggers and army deserters fleeing the war with Ecuador. The *brujo* accepts no money or creditocards—only trade for things he will find useful, or can trade on. What will Miguel give to him? The dish tonight—for there is only ever one dish on the menu, no choices—will be the *mutopargo*, an enormous multi-headed, many-finned mutant fish caught upstream in the poisoned Orinoco, where the chemical sprays that help the rainforest survive drain into the water. The fish are resilient and difficult to capture. When they are caught they often remain alive through days out of water as they are brought to the city. Miguel has heard that some are even still living as they arrive at the diner's table, to be eaten raw like Japanese sushi, a dozen eyes watching mournfully and fins flapping as the knife cuts home. Why would anyone want to partake of such a grotesque and cruel experience? Because the flesh of the *mutopargo* is the most delicious known to man. It is the food of the rainforest gods.

He stands and puffs out his shallow chest. She gets up and does the same—her plumage is much more impressive. She draws on her red spike heels and is taller than him by a head. He sucks in a breath.

"We will go to see the *brujo*," he tells her. Her eyes sparkle—she knows the story. He thrusts one hand into his pocket and closes his fist around the comforting keycard of the Ford Machos "Matador" Special Edition, and he knows now that he has her.

Nuevo Caracas, at night:

Black hardtop rolls by beneath the Ford Machos "Matador" Special Edition's fat tires. The car corners like a tram and Miguel holds the tiny electronic steering wheel with one casual hand. His other caresses Maria's naked brown thigh, revealed by the split of the red dress. So smooth, so smooth. She does not complain, nor does she agree to his touch. Ambivalence comes naturally to her.

Nuevo Caracas, at night:

Half the population is nocturnal. As the sun sets and the thunder clouds sweeping in from the forest fast-darken the sky, these creatures scuttle from their daylight boltholes to play or work, whether that work be selling their bodies (or body parts) or preaching infernal Papist damnation to anyone who will listen. The air is thick and damp and heavy with acidic ozone. Not that it bothers Miguel and Maria— the Ford Machos "Matador" Special Edition is equipped with an aerospatial-grade air conditioning system that keeps humidity, temperature and pollution levels within the car to acceptable levels. It is, perhaps, a little chilly. Miguel's nipples are erect beneath his brother's T-shirt. As he turns the corner where the *polizei* are threatening streetdwellers with electric batons, he decides he will see if Maria is similarly affected.

Nuevo Caracas, at night:

The city is a living organism, mutant child of the rainforest, an amoeba split in two, the squalid barrio and the cyberscrapers, with the streets the neutral ground where beggars and bankers can be murdered or raped or hustled by armoured riot-ready *polizei*, without fear of prejudice. Nuevo Caracas is nothing if not democratic. Bolivar would be proud.

Nuevo Caracas, at night:

The landscape changes as Miguel begins his ascent into the domain of the barrio. The Ford Machos "Matador" Special Edition's massively overpowered engine grumbles sulkily in its restraint mode under the sensuous haunch of the bonnet. Miguel's foot is barely touching the accelerator. He reluctantly forsakes Maria's delicate thigh and grips the wheel with both hands, begins to pay more attention to the road.

If a barrio gang emerges from a side street armed with a battering ram made from old crane parts, he will stamp his foot and the Ford Machos "Matador" Special Edition will—stylishly—carry them from nought to sixty in four seconds. The veloured bucket seats press their flesh reassuringly, ready to catch them if the Ford Machos "Matador" Special Edition rears like a stallion making a mad dash for freedom.

"How much further?" Maria whines, shifting in her seat.

Miguel is loath to take his eyes from the road. Street lighting disappeared a few miles back and the Ford Machos "Matador" Special Edition's powerful headlights spear the dark tunnel of the way ahead, picking out figures moving to either side. They pass a solitary *polizei-mobile*, parked on a junction with its doors and windows sealed, a single red light glowing weakly on its roof, like the last gas station for a hundred miles. They flash by at speed, into the dead heart of the barrio.

The barrio, at night:

In Nuevo Caracas, money changes hands and business is the order of day and night, the pursuit and accumulation of wealth, whether the vast riches of the interbankers or the savings of the whore hoping to escape the streets. Here in the barrio, there is only one business— the business of day-to-day survival.

The barrio, at night:

The city is alive but the barrio is dead, its heart ripped out by corruption and greed and man's inhumanity to man. Here life has been made cheap. A child can be sold for a meal. A man can be killed for a bottle of beer. When people have nothing, they have nothing to lose.

The barrio, at night:

Victim of the city, the barrio lies crushed between the cyberscraper and the mountains, compressed by need for that most valuable of commodities, real estate. A thousand people living in the space for a hundred with no power except for that which they might generate through ingenuity or desperation. A thousand people with dead hearts and dead minds and dead lives. A dead city to mirror its neighbour, so very much alive.

"How much farther?"

Miguel swallows hard and prepares to admit that he has no idea. The Ford Machos "Matador" Special Edition has slowed to a crawl and is making its way along steep winding streets choked with debris and the detritus of life in the barrio. Suspicious eyes view him from behind heavy hessian drapes and the paint-smeared windows of old trucks and buses that some of these people call home. He feels that if he stops moving, they will descend on the car like a plague of locusts and strip it of everything of value, including himself and Maria. In the barrio, everything has a value, including things that the city dwellers consider trash. That strange, sad thought terrifies Miguel.

The streets of the barrio are almost deserted, the thunder and lightning driving the people indoors to their shacks and shanties, to cling to their possessions in case the coming torrential rain tries to sweep them away. Miguel needs directions or they will circle this godforsaken place all night, and he knows Maria will not be impressed by *that*.

Instead, he thinks, she will be impressed by his nonchalance at stopping the car and asking one of the barrio residents for help. He has a small amount of currency in a billfold in his pocket, he knows that these people will want paying for information, and you cannot expect barrio dwellers to accept creditocards! He smiles at the thought as he parks the Ford Machos "Matador" Special Edition by a large black injection-moulded dumpster where the blue glow of a television screen seeps from the edges of a filthy sheet slung over the propped lid. Maria turns to him, horror on her face.

"Don't worry," he says. "I just need to ask the way."

And he gets out of the car.

Miguel is a foolish, ill-informed, spoilt youth of the cyberscraper culture. He knows no more of the barrio than the wild stories of the *brujo* and his ilk. He does not know that there is no reasoning with the people of the barrio when you have something that they want, especially when you are dressed in your older brother's best clothes and have a beautiful woman by your side. Maria could have told him this, if she were not paralysed with fear. Her family is less affluent than Miguel's—they live in the borderlands where you can smell the barrio, not just imagine it as a dark horizon or a scattering of twinkling fires in the distance. Maria has arrived home to find barrio kids in her room, rifling through her underwear drawer. She'd thought Miguel knew all this. She'd trusted him. Now, he has turned off the engine of the car, unlocked the door, and *got out.*

Miguel does not want Maria to know how frightened he is. He feels very vulnerable in his smart borrowed clothes as he approaches the dumpster. The side of the plastic box is painted in psychedelic swirls of luminous paint, cryptic symbols and figures. One resembles the Inca sunburst tattoo on his bare stomach. He fingers the tattoo self-consciously. He doesn't know what the tattoo symbolises, he just thought it looked cool. What if it offends a follower of one of the barrio's myriad religions?

He gingerly lifts the hessian that obscures the dumpster's lid. A child is inside, sitting crosslegged on a carpeted floor, leaning forward with its face pressed up to a television screen, nose almost touching it. Miguel cannot tell at first if it is a boy or a girl. The inside of the dumpster is strung with cheap Christmas tree lights, thin wires leading from them and the television up to the makeshift power lines sagging from the building next door. The supply is dangerous and unreliable and every few seconds the television picture recedes to a dot and then springs back again, each time apparently changing channel. The child does not seem bothered by this unintelligible assault on its senses.

"…*stay tuned stay tuned will you open the box or take the money buy the new CoCo-narcobar NOW it's full of flavour and can help prevent twenty types of cancer Ecuadorian paratroops landed today in northern sectors scandal hits new mining complex on the Orinoco delta stay tuned stay tuned stay tuned…*"

Miguel stubs his foot against the dumpster and the child jumps up, an enormous hunting knife in its small hands. Miguel sees now that it is a boy, no more than ten years old, with raggedy clothes, copper hoop earrings and a dirty face. The boy does a double take at Miguel's clothes, then waves the knife around in front of Miguel's nose. Miguel jumps back.

"What you want, clown man? You want my ass, huh? Not for sale, clown man. Go somewhere else! Unless you want to speak to my brother." The boy waves the knife again. Miguel presumes that this is his "brother".

"I'm not going to hurt you. I just want directions." Miguel holds up the palms of his hands in a placatory gesture. The boy's rodent eyes glance from empty hand to empty hand, and he relaxes slightly.

"I just want to watch my television, clown man. Why should I help you?"

Miguel keeps one hand outstretched and digs in the pocket of his brother's parachute pants with the other, bringing out the thick billfold. The boy's eyes light up like his television screen.

"What do you want to find in the barrio, then, clown man? Not much in the barrio to interest city types. Girls we got. You want girls, clown man?"

Miguel shakes his head and peels off several notes from the billfold. The boy licks his lips hungrily.

"The *brujo*, little man. We're looking for the restaurant of the *brujo*."

Standing up to his waist in the dumpster, the boy is a head above Miguel, and looks over his shoulder to the Ford Machos "Matador" Special Edition parked behind him. He can see Maria in the passenger seat. He reaches forward for the money but Miguel pulls it away and the boy almost topples over the plastic rim.

"Directions first. I'm in a hurry."

The boy clambers from the dumpster, his "brother" still in one hand. He wipes the other hand three times on his pant leg and offers it to Miguel, giving a slight bow as he does so. Miguel gingerly accepts the greeting.

"Bruno Del Santos El Rodriguez at your service," says the boy. "You want the restaurant of the *brujo*, and I am the man to show it you. But it is much too difficult to explain the way. I must come with you."

Miguel looks at the dirty boy and thinks of his beautiful car and the beautiful woman inside. He finds it difficult to imagine the two pictures in the same frame. But what are his choices? The boy has a big knife and information that he needs. He can only hope that Maria is not too appalled and that later he can clean any stains off the upholstery before his mother and father return.

"Okay," Miguel says. "But your brother stays here."

The Ford Machos "Matador" Special Edition roars around the barrio's narrow streets with greater confidence as Bruno leans forward from the narrow ledge of the car's back seat and points left and right, giving Miguel precise but, invariably, dangerously late directions. They are climbing into the storm sky, the ground above them thinning in its concentration of shacks and slums as they near the summit of the barrio.

Bruno is enjoying himself. He has never been in a car like this one and his seat gives him a perfect view down the girl's impressive cleavage. She slaps his face when he tries to touch her. She then spends the rest of the journey pressed against the trim panel of the passenger door, trying to get away from his hands and his pungent odour. He gives up and attempts to play with the myriad gadgets and screens on the car's dashboard, which earns him an equal rebuff from Miguel. He wishes he had his brother with him, then he would show these two some respect.

"There, there it is!" Bruno points at a skeletal ruin silhouetted above them against the lightning-torn sky. Miguel peers through the car's windscreen and uses the image to navigate his way through the last of the barrio streets, empty of life up here. Several dogs scatter from something large and dark they were gnawing in an alleyway. Miguel sees the steep, narrow road that will take them up to the mission church, and decides that he will drive the Ford Machos "Matador" Special Edition no farther. He parks at the bottom of the hill and switches off the engine.

"Good job, city man, yes?" Bruno grins with yellow teeth and holds out his dirty palm. Miguel smiles at him and they all get out of the car. Miguel fiddles with the keycard until he arms the car's defence system. A blue light glows softly on the dashboard.

"Good job, you pay me now, yes?" Bruno is insistent, urging. Miguel smiles again and presses a single crisp note into the boy's hand. Bruno looks at the note with disgust and spits on the ground.

"You promised me more, city man. We had a deal. You pay up, or I fetch my brother."

Miguel leans forward and gives Bruno a fierce push, sending the boy sprawling into the dirt. The note flutters away and Bruno chases after it on all fours, grabbing it before it disappears into the trash piled in the gutters. He stands and screams an obscenity at Miguel and Maria. Miguel picks up a crushed can and throws it at the boy, who runs away, cursing them in vivid language. Maria laughs, and Miguel smiles. He was worried that the evening was not going as planned, but now she seems genuinely impressed with him. He really showed that barrio kid who was boss man, didn't he?

The old mission church stands out above them, an unfleshed corpse of a building, a relic of a colonial past. Miguel takes Maria's hand and together they walk up the narrow street. In the gaps between the ruins they can see the inky expanse of the valley they have left, the city caught between the rainforest and this hard, dangerous place. Nuevo Caracas is a crystal ship afloat in a black sea. It seems so far away at this moment.

As they near the old mission, they can see fairy lights and candles flicker in the shell of the church, teased by the storm wind. Sheet lightning periodically turns night to day.

Miguel begins to feel nervous. What if this is just a myth? They have risked their lives—*and his father's car!*—to come here. Maria is hungry and impatient, and he has chosen to take her to the ultimate restaurant, which may or not exist, to eat an exotic—and quite possibly poisonous—mutant fish! Miguel, you are *mucho loco!*

Miguel guides Maria over the rubble-strewn courtyard. Big wooden gates lie forlorn to each side. Maria steps delicately and deliberately over the ground in her spike heels, allowing Miguel to steer her toward the softly illuminated plastic sheeting strung across the front of the mission. Maria stops and shrugs off Miguel's touch.

"There's nothing here!" she says petulantly. "This is no restaurant, it's just some barrio shack. Why have you brought me here, Miguel?"

"Come," says the *brujo*, stepping from the darkness, a slight figure in tapestry robes. "I've been expecting you."

Miguel and Maria freeze for a moment. The *brujo* is an old bearded man, not threatening in the least. Why should they be afraid of him? He smiles and beckons to them.

They follow the *brujo* without question. He sweeps aside the plastic sheeting and ushers them into his restaurant.

The interior is dark and smoky. Shadows chase shadows away from the glow of candles and lightning flashes. The restaurant is empty of customers. A dwarf waiter moves toward them with a glass pitcher of red wine.

The *brujo* shows them to their table—a packing crate covered by a cloth and marked with the logo of the Venezuelan Air Force, and two plastic picnic chairs. Maria graciously allows the *brujo* to pull the chair out for her before she gathers her dress around her and sits down. The *brujo* smiles toothlessly. The dwarf fills up their glasses with wine.

"You will, of course, be ordering our special," says the *brujo*, wringing his hands. It is a statement, not a question.

"Is it available?" asks Miguel coolly, raising an eyebrow.

"Of course!" says the *brujo*. "Otherwise, you would not be here."

The *brujo* and the dwarf disappear. Maria sips at her wine and looks around the restaurant, trying to see if they are really alone. Dark shapes flit around the periphery of her vision, but she sees no one. It is so hot in here, moisture rolling down the rippling plastic sheeting. A light sheen of sweat covers Maria's neck and shoulders, glistening in the candlelight.

Miguel's attention is fixed on Maria. He plays with the stem of his wine glass and tries to think of cool things he can say to her. Before he has a chance to deliver a stunning fusillade of compliments, the *brujo* returns, accompanied by the dwarf carrying a huge covered silver platter.

"Your *mutopargo*," says the *brujo*, and unveils the platter with a flourish.

The fish is still alive, shuffling ineffectively on the platter, surrounded by fruit and vegetables and Venezuelan *cachapa*, maize pancakes, and caraotas beans. It is a two-headed specimen, the most common mutation, and the two heads flip nervously in different directions, saucer eyes attempting to take in all threats. Multiple fins drum a beat on the metal dish. The dwarf places the platter carefully on the packing crates, then retreats with the *brujo*, bowing graciously.

Miguel and Maria and the *mutopargo* stare at each other for a long time in silence. The dwarf returns and gives them both sharp knives and forks.

"It's so beautiful," says Maria. "It's such a shame to kill it."

"Some people say it is already dead," says Miguel, testing the edge of the knife with his thumb. "It will have been out of the water for many days. It is just electricity making the fins and the head move."

Maria wants to believe him but the *mutopargo* looks at her mournfully, both heads swivelling toward her, as if appealing for feminine mercy.

"But it is supposed to be so good to eat," says Miguel, and makes a deep incision into the fish's flank. The fish shrieks and shudders. Miguel recoils and drops his knife. Maria licks her lips and picks up her own knife. She makes a bold, more positive incision deep into the fish's side, cutting a swath of white flesh. The *mutopargo* stops moving. Miguel watches, awestruck, as Maria slowly cuts the flesh on her plate and forks a piece into her mouth. She closes her eyes, chews and swallows.

"It is fantastic," she says. "It is the most fantastic thing I have ever tasted. Here, try some."

She cuts a swath for him and he accepts it from her. He rolls his eyes as he tastes it for the first time.

"Excellent," he says. "More. *More*."

He takes over and carves and feeds her, and in between takes pieces for himself. The *brujo* watches from the shadows, satisfied. He catches Miguel's eye and winks, then taps his palm pointedly. *What have you brought me in trade?*

Miguel freezes. How could he have been so forgetful? He reaches into his pocket and takes out two things—his creditocard and the keycard to his father's Ford Machos "Matador" Special Edition. The *brujo* accepts no cash. The Ford Machos "Matador" Special Edition…. His father's Ford Machos "Matador" Special Edition…. Is there no alternative? He looks across at the girl he has brought here. Surely not…?

Maria Del Fuego looks even more beautiful with her eyes closed in the ecstasy of exquisite taste. When she does open her eyes to see why Miguel has not fed her another morsel, he sees a look in her eyes which was not there before, a look that says, *Good work, Miguel. You've won. I want you.*

He smiles and forks another mouthful of food into her mouth. In the end, the decision is not so hard, after all.

Carlos will kill him, of course, but it is a small price to pay for such a wonderful evening, and maybe when he tells Carlos of his fantastic adventure and how he made love to the beautiful Maria Del Fuego back at their apartment, his brother will forgive him. They are heading home through the barrio at speed, and he turns down the air conditioning as it is getting cold in the car. He is, after all, wearing just his best silk boxer shorts, with his creditocard tucked safely in the tiny condom pocket, just behind the condom. He smiles at the thought of the *brujo* dressed in Carlos's clothes, an old man in the guise of a superhip Nuevo Caracas kid. Maria giggles at his nakedness but her laugh has a saucy edge, tinged by red wine and sexual tension. Miguel fights the impending erection as best he can. That would be *so* uncool.

Word travels fast in this urban jungle. These people have no need of internet or phone. The Ford Machos "Matador" Special Edition was tracked as it entered the barrio and allowed to pass through an elaborate series of gates and predetermined routes invisible to the eye of city dwellers. As Miguel climbed to the restaurant of the *brujo*, these gates and routes were closed behind him, and makeshift roadblocks sprung into place. On their way back down, Miguel and Maria are blissfully unaware that they are driving straight into a precisely prepared trap.

As they descend, Miguel begins to realise that they are not travelling the same route they took on the way in. The car slows to a crawl through streets that become tighter and narrower until he can barely maneuver the muscular vehicle. Belatedly, he knows he has taken one wrong turn too many. An old Cadillac, rusted and choked with foliage, blocks the road ahead. He looks into his rearview TV screen, preparing to reverse, and sees a party of figures appear out of the gloom. They are holding things in their hands, long things, sharp things. A sudden lightning flash illuminates them menacingly. Miguel utters a curse and guns the engine in a threatening manner, wheel-spinning and edging backwards, startling Maria who lets out a cry. The figures break ranks and Miguel prepares to get the hell out of there, but a dark shadow blocks the way. The bastards have towed a couple of wrecks in behind him, blocking his exit. He swears and thumps the steering wheel in frustration.

Maria has been watching the dim glow of the rearview screen. She stuffs a fist into her mouth and whimpers.

"What are we going to do, Miguel? What do they want?"

The Ford Machos "Matador" Special Edition will strike easily through either barricade but Miguel is worried about the paintwork on his father's brand new car. Emboldened by his encounter with Bruno the barrio boy and his successful negotiation of barter terms with the *brujo*, he decides to try to reason with them. Maria clutches her face as he gets out of the car.

Their faces are terrifying in the half-light. Ninety percent of barrio dwellers are Indian or *mestizo*, half-breeds. Their faces are painted in colourful chaos patterns. He suddenly remembers he is practically naked.

There are seven or eight of them. Behind them is a tractor attached by chains to the wrecks that were dragged to block Miguel's escape.

Miguel tries his best confident smile.

"Could you gentlemen please move your cars, and tell me the fastest way back to the city?"

The mob maintains a stony silence, then one nudges and whispers something to another, and they all fall around laughing and cackling. Miguel joins in, slightly relieved, but completely in the dark as to the joke.

"Bad night to be lost in the barrio, *caudillo*," says one of the men. "Storm coming. Barrio real bad place to be caught in a storm. All sorts of scum float to the surface."

Miguel laughs a nervous laugh. The barrio dweller calls him *caudillo*—it means strong man, or big man. It is, of course, meant sarcastically. Miguel realises he has made a terrible mistake. He starts to back away. The mob moves forward.

"Pretty lady in car."

"Pretty dress."

"Know how to treat a lady, *caudillo*?"

"That why you wear no clothes? You been playing fiesta with the lady, *caudillo*?"

"We show you how to play fiesta with pretty lady."

Miguel dashes for the sanctuary of the car. He decides to break out of here and to hell with the paintwork. These are not barrio kids, these men are evil *bandidos* who will beat him and rape him and leave him for dead. He places a hand on the door of his car, and his world explodes in blue fire.

Maria has slipped over into the driver's seat and armed the defences. The car's bodywork is now electrified and near-fatal to the touch of an intruder. The force of the shock has sent Miguel reeling into a nearby wall where he slams with bone-crunching force. He shakes stars from his eyes in time to see the barrio gang attempt a similar feat, undeterred by his own fate. They are propelled away from the car with all the sudden force of colliding magnets. Maria revs the engine and wheel-spins out of the confines of the alleyway, careening off the sides of the cars blocking the route and taking most of the paint off the Ford Machos 'Matador' Special Edition's right flank.

Miguel staggers after her, his bare feet slipping, and stubbing his toes in the trash-strewn alley. The barrio gang moan and wail in the alley behind him, the effects of the shock much greater on them with their metal weapons and studded clothes. If he is lucky, he can get away from them while they are still stunned.

Maria turns the car at an angle at the end of the road, preparing to make her escape into the wider street beyond. Miguel assumes she is waiting for him. He is wrong. The driver's side window slides down a few inches.

"It's been a lovely evening, Miguel," she says, blowing him a kiss. And she is gone, in a roar of over-powered American engine.

Miguel sinks to his knees in the middle of the alley. He has no car and no girl. He is lost and has no way of getting home. Maybe if he just waits here long enough, the barrio gang will recover and come and put him out of his misery. After all, what other choices does he have? How far is he going to get in a pair of boxer shorts?

"Hey, clown man, city man. Where's your lady, hey?"

Miguel winces. Bruno clambers from a nearby trash pile, dripping rubbish, his grin a yellow slash splitting his dirty face. He has his brother with him. He swaggers towards Miguel with the confidence of someone three times his age. It's easy to know when you've got the upper hand, even when you're ten years old.

"Want to know the way home, city man?"

"I have nothing to give you," Miguel says dejectedly. He is no longer even frightened of Bruno's blade. "I have nothing left."

Bruno closes one eye and peers at Miguel.

"Those are very nice shorts," Bruno says.

Well, thinks Miguel, as he stands on the brow of the hill with his city glistening like a teasing, out-of-reach jewel below him, *at least things cannot get much worse*. No car, no girl, no clothes. He may be naked but at least he knows his way back to Nuevo Caracas. If Bruno was telling the truth. He didn't like the way the boy had laughed as he ran away waving Miguel's silk boxer shorts in the air like an enemy flag captured in battle.

The storm, when it breaks, is like a cool relief, rain washing his body and the streets, the sky venting its anger on Miguel's behalf, as if the rich of the city are wealthy enough to bribe nature, and the barrio must pay the price. Monday morning and some very awkward explanations are a lifetime away.

For Miguel, it is going to be a long and interesting walk home.

DEATH, SHIT, LOVE, TRANSFIGURATION

BY BRIAN W. ALDISS

HERE'S HOW THE STORY BEGINS:

"God, it's crazed hot in here," Gyron said. "I'll toss some more wood on the fire." The gloom of it, the stink. The wayfaring hut was little bigger than an old wardrobe.

The flames blazed up, tinged with ice. Crackles of frost spurted into the room. About the beams overhead, ghosts of temperature played. The raven cried, "It's war! It's war!"

The other bodies drew slackly close, to stand with their backs to the freeze. Six of them, garbed in felt and leather, booted, shoulders padded with musk ox pelt. They burnt no lamps; it was dark enough. Their eyes were blinkered with plastic clamps. Their mouths were fleshed over with dermoflesh. They had not eaten for forty-nine hours, brother or sister, and were fattening on starvation. Gyron fed them by clyster occasionally on the long journey.

They smelt of smouldering horsehair.

Personal hooks were slung over the ceiling beams. They hung themselves there, hung like hams, jostling the festooned bats. Hot or not, they slept. Sleep was now a compulsory programmed six hours. After the first half of the sleep programme, they came down and assembled outside the hut. There the president's body, bone and gristle, lay in its box. The box was decorated with a wreath of lead-encased dog skulls and chicken legs. Through the coffin's oval window, the president's face looked blindly up towards vacancy. His hair continued to grow. It patched him whiskery white and black like a badger, as if he was having a last joke on the world of men and clones.

Chief Attorney Gyron scrutinised his compass, seeking magnetic south before giving his orders by high-decibel whistle. The clones shouldered the coffin. They stood mute, disconsolate, undreaming, eyes smouldering in blind mode. Gyron flashed the stinger at their backs to get them moving. They marched in ashen step. Gyron climbed into his walker and followed; its metal legs going clarp-clarp-clarp, like a psychotic band playing yakkahula.

The Great Grave was half a day's hike away—if they could reach it safely. There the dead president would be liquefied, turned into earth-libation.

The eggs of day unscrambled. The Six Blinders glowed their red bars overhead in a pitch-black sky. No one heard their ringing.

Chainganging onward, the clones in their skulls heard only the music of Happydrug. They rejoiced in their misery.

Less than three verts from the sleepwalkers, an enemy force was converging on them, foul of mouth and body and mind, intent on seizing that box containing the decomposing figure.

The singing sleet, the driving snow, the knife-edge mountains, the way forever ill-defined, the enemy soldiers wrapped in rags like mummies, their faces lined, veined, veiled like hags. Moving as if dead along narrow defiles, their old wounds opened and bled steam and shit. All, all bleak, open to frostbite, stones, cross-winds, ice. No comfort in nature and none in their minds. Their paralysed king was hauled along in a bamboo cage, up and down precipices. He screamed occasionally on a weak feminine note.

Above the frowsty be-furred heads, banners waved, lugged with great effort against the streaming screaming air. These were the insignia of religion, lending their unction to the death patrol. Monks, heavy of crux and coat, laboured up the cliffs with coffers containing sacred relics of medieval kings, Pochavan, Destriney, Gravelbig, Gray the Ungrateful, Evalsh, now little more than gristle…. A foot here, a femur, shards of bone, a slice of pate there, parts unknown, bound in silver. Faith, institution, nought if not nation, saved over centuries. Fought over, prayed over, hoarded in times of destitution, furtively kissed. Bringing the bearers no good.

But where the relics were, there ran tribal blood, thicker than serpent ichor.

Soon now the ambush, swift attack, thrust of sword, death in a stained storm. Could they but seize upon the old president's corse, no less!…. Maybe quantum basis of consciousness, brain's season, could be restored….

Unreasonably, they dreamed of reason.

Bedevilled, they closed on the president's procession with every step and slip. As a spur, the introit oft-repeated from frost-encrusted lips:

"On, brave shaggers, together, alone! One more dose of death that we may have life.

On, shaggers brave, brave to the bone! The thrill of the kill that we may have life.

"Life, Meat, and Frig…"

Not, intoned the monks, Life Everlasting. Nothing so big.

"Simply a moment of light, a lick of it to the heart, a finger up the rotten boghole of Time. Pain, pah, grasp the prick of it—that something forgotten as a fart might start up again."

Their tongues hanging out, panting like pye-dogs among the granites and scree. Minor-key stuff, every word snatched away into the howling gloom. Knowing nothing of the functionality of futurity, yet they were the very switchback of history itself.

Now, high above their tawdry backs, the Six Blinders crossed. Blanket lightning turned the sky white as lady's lace. Over the eyes of the ferocious marchers, nictitating membranes oozed into place. The count of twenty, then the phenomenon was past; all that prevailed was the starless black above, crossed by its dull red spirals.

"One more pass, doom-drinkers, womb-shrinkers, one more abyss, then, Hell's snatch!—dogs' piss, weird powers, and down in a batch upon the clone stinkers! Then back to the towers of home, lustily, the president ours, in our custody, bony in his lone bower!"

Dreamless, the six numb clones march onwards with their mute burden. Left right left right, moonlight in their calmed cerebra. Heaven is hailing down at them. The attorney goading them, they move through a valley bottom. A waterfall rushes from the ground, splashes against bare rock, bursts upwards to the peaks concealed in mist. The noise is a death rattle in its own throat.

Gyron knew that the Great Grave was not far now. There the body of the old president would be dissolved in acid and poured into the barren soil. An ash tree would sprout from his remains. Presidents had an easy life, sheltered from birth. The symbolism of their last journey into the harsh lands appealed greatly to his mind. All was symbolism. He understood that now. In this world, reality was what he could never possess, grasp though he might at its trappings. He was himself a symbol of that greater thing which lay beyond, like a figure on an embroidered screen, crowded with other figures.

And when his party had gained the place of the Great Grave, he too would enter it with his president.

He rejoiced, and the sleet slashed down upon his shoulders, its blades on his blades.

Kirpal Quarmless was born in Singapore. The Quarmless Theory was born when he was in exile in Europe. Quarmless investigated the nature of consciousness. He was curious to discover why the minute quantum level of particles, waves, and atoms hardly seemed to fit in with the classical picture of a unitary, Newtonian universe. Quarmless's early mathematical probings investigated this discrepancy. Early on, he saw that mathematics itself—embedded in the old systems—was no more than a sophisticated tool of the discrepancy. A truthful view of the world was not to be gained through the faulty lens of its prevailing math.

His findings were that the old classical theory of the universe, together with that of small reactions, was by its nature unable to explain the means by which consciousness had developed. His paper, *Multi-World Understanding of the Quantum* (2050), pointed the way to a new mathematics, beyond mere aggregates of computations—beyond which aggregative system even the most powerful computers had proved incapable of progressing.

Quarmless's proof that computation had led to a perceptual flaw in all scientific method was both difficult to comprehend and highly challenging. It overturned the accepted nature of the world. Finding his name vilified, Quarmless went into hiding.

Unpalatable scientific findings in the past had eventually led to new levels of understanding. So it had been with Darwin and with Einstein. Now it was Quarmless's turn. There were open minds who saw—with hope and trepidation—that he had opened the way to a revolution in thought. The universe had to be reinvented.

When the early heat of denial had died down, it became apparent to mathematicians and scientists, most notably the Cochrane group working on MFD, Mental Functional Distortion, where that revolution in thought lay. Earlier attempts to explain the workings of the human brain in terms of probabilistic physical activity were dead as the rhino. The old precise mathematical laws on which the twentieth-century computer had operated were insufficient to explain the operations of consciousness. Certainly, those laws (man-made and not natural) had proved unable to explain a universe which necessitated consciousness. Pain, love, the appreciation of beauty, the delights of music, a thousand sensibilities owing no tribute to reason—these and other effects were explicable only in terms of a functionality beyond classical and quantum parameters. Quarmless's *multi-world math* added a dimension to what had hitherto been the equivalent of Flat Earth thought.

"I'll change the world," Quarmless said, sheltering, growing a disguising moustache in a Bavarian village. "Descartes, Aristotle, changed mankind's perception of the world. So will I."

He was bitter. He saw his tremendous idea slipping from his fingers. Others would develop what he had started. He was nineteen. The *syonanto*, the Singapore secret police, were after him.

In search of new quarmless integers, the Oxford-based Cochrane nexus of scientist-philosophers began to perceive practical applications of MW math such as Kirpal Quarmless himself had never considered.

"In five years, we shall hold within our hands a new understanding of reality," announced Robert Penstimmon, the leading member of the group, in a perivision broadcast.

It took them not five years but four. And the suicide of one member of the group.

They found themselves staring into a reality which had barely been dreamed of.

The quantum theory had offered an explanation for the nature of chemistry, the existence of planets, the phenomenon of heat and cold, and other familiar properties. Sentient beings such as humans had seemed to exist—had thought of themselves as existing—in a classical world. Yet *consciousness*, that mysterious quality, had originated on quantum levels. The new MW math showed mind to be a factor of something beyond, developing in what came to be called the quarmless universe.

"What will we find in this so-called quarmless universe?" an early interviewer had asked Penstimmon.

Penstimmon's reply became world-famous. "Death, shit, love, transfiguration…"

He eventually explained that, in layman's terms, humanity was confronted by an entirely new perceptual breakthrough. No longer was brain conceived of as a function of algorithms projected from classical structures or quantum functions. It was rather the instrument of metamorphosis into another frame of being.

As some acceptance of the new concepts dawned, there was a brief period when stock exchanges slumped, armies drew back, and governments hesitated to raise taxes.

"We have long recognised that the phenomenon of consciousness is strangely at odds with the known universe," Penstimmon said in a television interview. "We have mistakenly perceived it as a contradiction, a unique factor. It is rooted in the laws of Quarmless Theory which are the actual motivators of our universe. Only now can we actually formulate how our world operates. "

"So there's another universe we can see into? Can we—well, can we get into it?" asked the interviewer.

"The question is one of new perception."

"So this quarmless universe isn't real? You can't walk in it?"

"You must put the case the other way round," Penstimmon said. "We should be *there*. We really are there. We have all been walking about *here* all our lives, within an illusion. Much like being trapped within a bubble. Ours is an unreal perceptual universe."

The interviewer permitted herself a wintry smile. "Countless millions of people are going to find that hard to believe, professor.'

"Countless millions of people still find it hard to believe the Earth goes round the Sun, yet it is so."

"One last question before we go over to the Sports desk. Is this new place going to be heaven or hell?"

"We have all had to live in both Heaven and Hell, in this perceptual bubble of a universe. When we escape all illusion, the quarmless place is bound to be either one or the other. As yet, it is too early to say which it will be."

END

✦

This story is, as far as I know, unique in the history of fiction. It was written over several generations, constantly revised, updated, never finished.

As I have related elsewhere, an ancestor on my mother's side was a Chinese of humble origins. He lived as a simple indentured labourer in Capetown, and was moved about Cape Province at the will of his master, Ben Moshe Joel. When the Boer War broke out, Joel presented him as a cook to the British 4th Hussars. It was a patriotic gesture.

My ancestor's name was Jiang Office, or so it had been entered in the records of the Joel Exportation Company. Jiang's expert cooking made him a favourite with the Hussars, who corrupted his family name to the English first name of John. John he was, and John he remained. During the battle of Omdurman in 1898, John saved the life of a young English officer, by name Winston Leonard Churchill. In gratitude, Churchill freed John Office, buying him from the regiment and taking him back to England as a personal servant.

The erratic Churchill soon grew tired of John's limited English. He locked the man up in a stable block and gave him four weeks in which to write a piece of prose of not less than one thousand words in perfect English.

John failed to do this, and was sacked from Churchill's service. He took with him the prose piece he had written, entitled "Trafficration" (his English still falling short of perfection). He found his way to London. There, a poor Irish family in London's East End took pity on him, giving him a floor to sleep on.

The prose piece John wrote in Churchill's stable marks the beginning of a story entitled "Transfiguration." It was a tale of the suffering and exodus of a whole people. Setting the piece to one side, John proceeded to teach himself idiomatic English. Perhaps he hoped that by so doing he might again find a place in Churchill's employ. Instead, he fell in love with an Irish girl who had a room in the same lodging house as he. Her name was Rosie Mulvihill. Her father, Pat, was a brutal drunkard from whom she had escaped.

Unfortunately, Pat caught up with his daughter and hauled her back to Ireland. John was heartbroken.

Winston Churchill became newsworthy at this time. He had been a Conservative MP. His switch to the Liberal Party created a scandal. The sensation was fully reported in the *Daily Mail*. It gave John an inspiration. He wrote an article entitled, "How I Saved Churchill's Life," which was immediately accepted by the newspaper.

The article was published under the name John Orldiss. On the proceeds of his sale, John could afford to sail to Ireland in search of Rosie Mulvihill.

She was easily found. Pat, her father, had murdered her mother in a drunken fit, and was up for trial. It was in all the papers.

John married Rosie, living on her slender earnings as a singer while he tried to break into journalism. They produced three children in quick succession, Brendan, Boyce, and Bryan. Only Bryan survived infancy.

John sometimes acted as messenger for a man whose business he never knew, who went by the code name *Bugle*. He also turned again to revising his first prose piece. To him the story was talismanic, a sign of his freedom. He now diverted the theme of exodus into an account of armed struggle between two gangs of bandits in the mountains of western China. He had read a brief outline of an actual encounter in a South African paper.

Although he never managed to complete the story, he wrote other ephemeral pieces of fiction, some of which were published in ephemeral Irish magazines.

However, luck came his way, for Rosie's singing in a Dublin pub attracted the attention of James Joyce, the poet and novelist. Rosie and John became quite friendly with Jim before he went abroad.

John made his son Bryan run messages for Bugle. Bryan disliked the man who called himself Bugle, and he disliked the job. Once when he had an urgent message with which he was to run to deliver to a butcher's shop, he threw the note into a waste bin instead. Two days later, Bugle was shot dead behind an hoarding advertising Nivea Vanishing Cream.

John arranged for young Bryan to sail over to Europe with the Joyces. Jim got fed up with Bryan and left him in Paris with a man called Jacques Latourette, a Belgian who had lost a leg in the first world war. Latourette had been a sapper; he possessed considerable scientific knowledge, which he perverted into distorted cosmological theorising. His theory that Earth was a satellite of the Moon, rather than vice versa, won many adherents world-wide.

The Belgian was slightly unhinged. He could talk of little but the war and what he and his friends had suffered on the western front. He had secured a considerable advance from the French publisher Gallimard for his story (and was in any case a man of independent means, keeping as he did two mistresses and a parrot called Colette). After the first sentence of his account, "'God, it's crazy cold in this trench,' Desrolles said, as he and his soldier mates huddled together on the duckboard," Latourette ran against a writer's block which had lasted six years. When young Bryan Orldis (as his name was given on his passport) came along, Latourette took a gamble.

"It's a wonderful story, but I don't know how to proceed with it," said Latourette. He chewed his fingernails.

Bryan, immature as he was, had taken a liking to Fifi Fevertrees, the Belgian's Anglo-French mistress.

"I could write the story for you," he said.

He got the job. He lived in Latourette's house, where Fifi soon taught him to, as she put it in her genteel English way, "kiss her between the legs."

One day, a letter came from Ireland. It was from one of Rosie's old aunts. She wrote to say that Rosie had died of consumption. John, distraught with grief, had tied the tie he wore to her funeral round the handlebars of his bike and cycled at full tilt into the Liffey, where he drowned. He had left nothing of value. There was only an unfinished story, "Transfiguration," which he begged his son Bryan to finish, since it represented freedom to him. Rosie's mum enclosed John's note, together with the tattered manuscript.

Shortly after that, Bryan was badly beaten up in the street one night. Breeze blocks were dropped on his knees, his arms were broken with an iron bar. "That's for Bugle," someone said, vanishing into the night.

Bryan finished neither Latourette's novel nor his father's story. He tinkered occasionally with the story, removing the Chinese flavour, since he wished to conceal his own ancestry. He inserted instead some of Latourette's far-fetched scientific theories.

He and Fifi were married. He remained a cripple for life. Fifi took him to live in Berlin, where her uncle Desmond Miller (or Muller, for accounts differ) played violin in the orchestra under Furtwangler. The married couple quarrelled violently. Bryan was always in pain. Nevertheless, Fifi bore him three children, two girls and a boy. Bryan ran a small radio shop in a street off the Kurfursterdam.

When the Nazis came to power, he was in trouble. His oriental looks, hardly discernible to most people, marked him out as non-Aryan. Fifi was warned she should divorce him for her own safety. She showed spirit, defied the Eugenics Office, and stuck by him. Uncle Desmond persuaded Furtwangler to speak for his crippled nephew-by-marriage.

Meanwhile, one of the daughters, Claire, showed literary talent. She obtained work in the office of a fabrics factory, soon becoming manager of a small branch in the suburbs. Here she typed out "Transfiguration," converting it from ragged holograph to orderly typescript.

Claire had a hatred of military activity. When her factory was forced to divert to making Wehrmacht uniforms, she rewrote the "Transfiguration" story, making it an engaging tale of two tribes of bunnies, one lot white, one lot brown, who first quarrelled and then made peace. They were going to live happily ever after, when Hitler's armies poured into Poland. War broke out. Once more the story was never finished.

One night, her crippled father disappeared. They never had word of him again. They did not doubt that the Gestapo had arrested him. With the connivance of Uncle Desmond, Fifi manged to get a train to Paris, taking the children with her.

Fifi still retained a certain charm. An American journalist working in France, by name Raymond Gram Swing, took a less than fatherly interest in her. They became lovers, living in Montmartre. Fifi's son died of cholera. Meanwhile, the Nazi armies were sweeping westwards, towards Paris. America was still neutral. Swing pulled strings at the American Embassy and provided Fifi with a forged British passport. He flew out with her and her two girls forty-eight hours before France fell and the Wehrmacht marched into Paris.

Fifi was distraught. Her English was poor. London was looking particularly bleak as it prepared for war. She realised that Swing was not for her; he seemed already to have lost interest, being caught up in the British cause. She thought of rejoining Latourette, but Belgium, like France, was now enemy territory.

A colonel in the REME invited Fifi to spend a leave with him in York. They stopped for a night in Peterborough on the way north. It was here that Fifi's daughter Claire was kidnapped by a Norfolk farmer.

With Claire, struggling and kicking as she was shoved into the back of a Volvo estate wagon, went her overnight bag, containing the various abortive drafts of her grandfather's story.

Wilf Deacon, the farmer, never received the ransom he demanded for Claire. Indeed, under wartime conditions, the demand never reached Fifi, who by now was living in a Scottish castle on the shores of Loch Awe, waited on hand and foot by Penelope, the REME colonel's adoring mother. The colonel did not survive Dunkirk. His mother and Fifi lived quietly together beside a lucrative trout stream. Fifi had found happiness at last. Penelope died at the time of the Korean War.

After beating and maltreating Claire, Wilf Deacon accepted her as one of his daughters (who were also beaten and maltreated)— rather an unusual acceptance, since she soon bore him a child. Their relationship had been consummated in the warmth and comfort of the pigsty. The child, a daughter, was named Kate after Wilf's favourite dog, a Rottweiler bitch. When old enough, Kate attended the local school under the name of Kate Aldiss.

Wilf had spent fourteen months on Venus, as he claimed, and was something of a local celebrity. His article in the local paper, *The Wells Chronicle*, told how Venus was sunny but very cold. Farming conditions were difficult, owing to the fact that there was no soil. But Wilf had taught the locals how to grow potatoes in rock. As a result, he had been crowned King of Venus (or Rock, as the locals called it).

When kidnapping Claire, he had intended to take her back to Venus with him. As he explained, the Venusian women had breasts but very little else.

Stimulated by Wilf's stories and her mother's history, Kate began writing in earnest when still a child. She was determined to complete "Transfiguration" as a novel. Her upbringing had made her tough-minded. Such was her tragic vision, she concluded that humankind could exist only in a state of war. She strove to tell a tale full of death and shit, with a slight love interest running through it. Filled with the beauty of her imagination, she attempted to write it as a verse epic.

Her desire was to encompass a whole imaginary world, filled with science, religion, glory, dedication and—in the words of an earlier version of her inherited story—death, shit, love, and transfiguration.

The war ended with the defeat of Nazi Germany.

Kate's epic was well launched when catastrophe struck. A squadron of the Luftwaffe based on the remote Lofoten Island of Minhin had never accepted defeat or the death of Adolf Hitler. The brave fliers there believed that a sudden strike against *Gross Britannien* could reverse the course of events. Hermann Goering had been mistaken in his attack on British cities; what was needed was a strike against British agriculture, and the wretched little island's economy would collapse. (Such quasi-scientific theories were common in those days.)

The last brave flight of the Third Reich took place at dawn one morning on the anniversary of Hitler's birthday, 1955.

A Dornier, flying over Wells in North Norfolk, unloaded its clutch of bombs, inscribed variously **HEIL HITLER, FOR YOU, TOMMY!** and **THE WAR IS OVER!** The clutch fell on Wilf Deacon's farm, killing three goats, a ancient Rottweiler called Kate, Wilf himself, Claire herself, the very pigsty in which their love had been consummated, and an incubator full of newly hatched chicks. Kate had to abandon her manuscript and find a way of supporting herself.

❖

Here I must inject a personal note. Kate Aldiss was my great-aunt. She married a man called Alastair Holman, spent thirty-three years working as secretary for a solicitor (Sims & Malpractice) in Fakenham, and never took up her pen again. She took up gardening instead.

It was her son, Charles Holman Aldiss, who was next to attempt to complete "Transfiguration." Charles, becoming a regular soldier, had plenty of leisure time. He sent an outline of his story to the publishing firm of Faber & Faber, and was commissioned to write a novel with that title.

Charles was not a natural writer. He attended a Creative Writing course in Norwich, as a result of which his idea was to concoct a huge bestseller. He knew that SF was popular, having been so informed by a neighbour, a Mr Leonard Fanthrope. His intention was to transpose the story into the future, setting it in the Twenty-First Century, which he visualised as rather rainy.

He took a hint from earlier versions of the story. Social life had deteriorated steeply, owing to rock 'n' roll; the presidency of the United States of America had become a hereditary office. Based on this premise, he began to plot a trilogy. Faber did not complain, and neither did Faber.

Unfortunately for Charles and literature, the Falklands War broke out at this juncture, and Charles was sent to those remote islands in the South Atlantic. There he was killed by an Argentinian-launched French Exocet missile in the very hour of his arrival.

Knowing that I occasionally dabbled in the literary arts, Charles's executors sent me the worked-over drafts of "Transfiguration," with a request that I should complete this fascinating story if possible, "for the continued honour of your family," as they expressed it.

The manuscript was forwarded to me at the Holiday Inn in Kuwait, where I was holidaying with my wife. (Her cousin Sandy worked in the British Embassy there.) This was 1990. We had to leave in rather a hurry on the last plane out before Saddam Hussein's forces occupied the city. In my haste, I left the precious manuscript bundle behind. No doubt it now resides in a Baghdad museum.

I concealed my grief at this carelessness as best I could, saying no word to my wife. Only this year did she think to mention to me that she had in her possession a photocopy of "Transfiguration" in all its forms. She had prevailed upon her cousin Sandy to copy the MS at the Embassy, so that she could study it herself. (She made little sense of it, incidentally.)

Great was my joy and relief!

Now I am planning to reshape the story myself.

I shall not allow the bomb outrages in London to distract me.

By way of life insurance, I publish above the story as written and rewritten by four generations. With a few odd touches of my own, here and there.

THE PACT

BY ANDREW STEPHENSON

Giving his helicopter its final instructions, Sir Henry Gypter stepped clear and watched it lift into the sky.

The fierce downdraught probed and tugged at his clothing. A sense of nakedness hit him, a recognition of being on enemy soil, encircled by a wilderness that reached, not-so-symbolically, into seeming infinities of inimical nothingness.

His hand case felt tiny, too light, empty. Were all special needs covered for the next two days? Everything had to be in the case, or in his pockets. No hope of local supplies.

Then so be it, he decided, and turned to confront the rococo absurdity of his hostess' home.

Ahead sprawled flawless lawns, an emerald bib for the broad sandstone frontage. Ranks of windows ignored him. From out of the mansion's maw a tongue of rosy marble steps dipped into white gravel, as though lapping milk. Shadowy within an entrance arch, ebony doors stood shut against the world.

Behind those modest portals lived This Week's Sucker.

The deal clicked back into perspective and felt good again.

He began a leisurely stroll across the grass. To encourage in himself a formal frame of mind, he hummed one of the jollier passages from Liszt's *Dead March*, reasoning that very likely his doubts had been appropriate. Even robbing a baby of its candy could be hazardous: the brute was liable to slobber on you.

Before he had quite attained the steps, some subspecies of uniformed servant bolted from cover and tried to deprive him of the case. The fellow seemed put out on being invited to desist.

Another footman, his expression the merest hint removed from a sneer, deigned to open the doors. He was kept waiting, while Gypter awarded the antique brass hinges an unhurried appraisal.

"No automatics?" Gypter enquired, eventually.

This one maintained a professional cool. "None, sir," he intoned. "The mistress deems them to be inconsistent with the traditional character of the House."

"Hah," said Gypter and swept through into the cavern of the entrance hall. *Soon fix that if I buy in,* he decided. *Could use such shortcomings as a lever to drop the price—*

He gaped in stark disbelief.

Maxwell's ghost! What a heap!

Hectares of shiny wood loomed on all sides. Warehousefulls of chairs and sofas, bulging with PVC and leather, crowded close, tempting travellers to entrust any weariness to their comforting embrace. Antique knick-knacks, fashioned from scarce materials, perched in parasitical flocks upon herds of lesser furnishings. Antimacassars and beaded covers abounded. Left and right, two stunningly gruesome gilt-framed paintings, of a Dionysian revel and three winsome Victorian brats hugging fluffy kittens, spanned the walls and challenged the sensibilities. Directly ahead, a veritable Jacob's Ladder of a stairway, corseted in pink-motif Bokhara and ribbed by varnished brass rods, zigzagged high into dark mahogany heavens, whence constellations of crystalline lamps shed a sparkling golden glow.

At the foot of the stairs his hostess waited. Tight mouth smiling. Eyes wary.

A glance matched mistress and mansion, in age and style.

"Sir Henry," she rustled, extending a hand. "Most welcome."

Setting his case on lustrous parquet, Gypter took her hand. Its skin was dry-smooth, reminding him of snakes. Her antiquated tie-dyed clothing reminded him of an accident in a paint factory.

Watch your back, he told himself. *Patricia Amelia Lourat did not get rich through sheer niceness, no more than you did.*

"Yah, thanks," he said. "Good notion, Pat, asking me over to size up the place. And it's Henry to friends."

The acquisitive flunky was making moves on the case again, so Gypter paused to shoo him off and resume a firm grip. He made a performance of unbuttoning his overcoat.

"I expect you are tired," said the old lady. "Swanson will guide you to your apartment. When you feel ready to join us, my other guests and I will be in the southwest drawing room."

The servant glided partway up the stairs and paused. Before Gypter could follow, Lourat whispered, "Dear Swanson would be much happier if you permitted him to carry your baggage. Otherwise he feels slighted."

Gypter eyed the flunky, then beamed at Lourat.

"That right? Okay, if he's a really good boy, maybe soon."

Lourat blinked. "Henry," she said, "please consider that our bargain is not yet confirmed."

Beckoning to Swanson, Gypter proffered the case. "Must be your lucky day." He favoured Lourat with his sincerest predatory grin. "Tell the gang I'll be right on down."

There was too much talking and bonhomie. Fully five minutes were wasted before he could entice Lourat aside.

Her other guests had proved to be stockholding nebbishes or senior employees, mingled with a sprinkling of ecofreaks on their best behaviour. Ostensibly it was a relaxed social event, to let prospective partners get acquainted.

But Gypter knew who was under scrutiny. Ineptly concealed reactions and intemperate remarks suggested he was hardly flavour of the month with some present, particularly not the 'freaks.

Sir Henry Gypter, it appeared, had a Reputation.

"Listen up, Pat," he began, once they were safely cloistered in her study, "a lot of bull's been spread around about me."

"Surely you mean 'synthetic fertiliser'?" she murmured from the depths of an easy chair.

He managed a chuckle. "Hey, not all of my businesses depend on clever chemistry and ingenious engineering."

"Indeed not. Sagacious shysters are said to do their bit."

"The world seethes with small minds and jealous hearts."

"Is that it?"

He assumed his placatory simper. "Oh, I make mistakes. Now and then. But I try to square things with any who get bruised."

"Yes." There was a long pause. "So I heard. A man who can accommodate conflicting viewpoints and unexpected realities… Which is why I thought you might be worthy of an approach."

At last, the nub of it.

He waited. And had the good grace not to smirk visibly.

It was long tale: of a business founded on early successes and happy chances; of ethical practice winning approval from a public sick of cynical corporate greed; of canny altruism inspiring the Lourat Association to make several now-famous wise investments.

It was also a tale with a down side: of public complacency allowing bad old habits of industry to creep back; of shifting fashions; of crucial mistakes dragging the Association towards the verge of ruin.

All in all, it was a pathetic and whingey sort of tale.

Gypter had sat through many such, however, and had learned how to look interested, even sympathetic, while his brain figured what might be in the ever-so-sad situation for him.

"By now," concluded Lourat, "the wealth of the Association is, in effect, reduced to a single asset."

Gypter raised an eyebrow in feigned incomprehension.

She looked disappointed.

"Okay," he admitted. "This estate."

"You, on the other hand, possess wealth and influence enough to preserve both it and its essential character."

She waited for him to infer the implications, before adding:

"We thought you might appreciate a chance at first refusal."

He sat up. Was he awake? They were discussing an irregular splash of prime, ripe, unexploited temperate territory. Verdant mountains and flatlands stretching four hundred and something kilometres east-west, a shade over three hundred north-south. A fabulous treasure of clean water and topsoil, timber, minerals, construction space, transit rights, gene pools to plunder…

"Henry, you're drooling."

Embarrassed, Gypter wiped his mouth, found it dry, looked up at Lourat. She snorted.

"My advisors were not wrong, it seems. Let's not beat about the bush. This place needs your money. You are someone who can, properly stimulated, make quasi-civilised behaviour pay handsome dividends. Very well. You want control. We want survival. The makings of a partnership?"

Her eyes, gem clear within the papery wrinkles of her face, pinned his.

"Hell," he said, "you know you've got me."

"Nevertheless, think about it." She stood up, surprisingly briskly. "I anticipate you will find our terms agreeable. The documents will be brought to your apartment at noon tomorrow. Before then, why not explore the grounds? Morning would best. Swanson can equip you with an aircycle and other necessities."

A walk in the woods? Repulsive thought. No hardways, not anywhere. Just squishy, sticky mud. Branches and thorns to snag clothes, gash skin. Rabid animals to launch frenzied attacks.

"Must I?"

"It's quite safe. And will clarify things. Please."

What an asset-stripper must do for a profit these days, he thought, nodding as if it seemed an exquisite idea after all.

For an experience, the outing began ordinarily enough. Swanson knocked on his door an hour before dawn to deliver an armful of "suitable apparel." This was military-style boots, trousers and field jacket. All, Gypter was pleased to see, manufactured by one of his own companies.

"Nothing but the best," he quipped to Swanson.

The footman inclined his head minutely. "The mistress felt, sir, that you would be happiest using equipment in which you had most confidence. Moreover, should it fail to satisfy, you would be in a position to apportion blame effectively."

Breakfast was experienced in a vastly under-occupied formal chamber. Gypter's clumsily massive dining chair faced along a table whose far end seemed to lie beyond the horizon. Somewhere down there, he was sure, sat a Citizen Kane look-alike, similarly engaged in a joyless repast. Meanwhile, at the near end, Swanson lurked randomly, inflicting occasional personal services.

He could not quite nail down what it was about the footman he was learning to loathe. He had never taken to the Wodehouse stories starring that super-butler, Jeeves. Perhaps it was the way Swanson attained a similar perfection without evident effort, and without having to humble himself in the slightest, which so irked. Or perhaps it was Swanson's seeming indifference to what Gypter thought of the service, so long as he permitted Swanson to provide it. Damn fellow was scarcely human.

With the sun at last clear of the eastern mountains, Gypter was led out to a sprawling fieldstone building roofed in mossy slate. Wide wooden doors swung open as they approached. Ceiling panels shone down upon assorted one- and two-person vehicles.

"The mistress suggests you take a two-seater aircycle, sir," said Swanson, indicating a spotless machine by the doors. "The handling characteristics are superior. And you may find the additional capacity convenient."

"Whatever for?" Gypter bent to look inside. The autopilot was a top model. Again, one of his own. Gadgets galore littered the interior with pretty detailing, including handy touchtabs for electronic window blanking. Sumptuous craftsmanship complemented a design that was spacious almost to a fault, the wide fold-flat seats together virtually amounting to a bed.

"One never can predict," the footman explained. He released the driver's door and held it wide.

Well, if there was something about the machine meant to harm him, dodging out now would be futile. He was here for a full day yet. They could get him later, if obliged to.

He settled inside and began the startup procedure.

There were no difficulties. The aircycle performed flawlessly. Flying high above the forest, snug within the cockpit canopy, he savoured the purity of the scenery and its precious virginity. *Money-money, money-money*, muttered the engine behind him. Such awesome potential, if he could but convince Pat Lourat and her cronies of his honourable intentions. Thereafter, the sack of Rome would pale into a sincere attempt at civic conservation.

He flew far enough to be out of sight of the mansion. That alone called for a journey of several kilometres, as the pile squatted on a hilltop, visible from—and hence viewing—far and wide. Then he needed to find a spot where it was possible to dismount without at once being assaulted by rampant wilderness.

The map showed a view point, with proper landing pad, beyond the first ridge. A rocky cup in the hillside provided screening for a bit of clandestine business. He settled there.

As an aircycle's autopilot was not automentive, it could be trusted to take no interest in events not concerning the vehicle. But the chassis might be bugged; playing safe, he moved away downhill until bushes intervened.

From a side pocket he slid a long, flat box. Outwardly this was plastic tat, textured cunningly to seem cased by real snake skin. Gullible tourists snapped up such goods in furtive markets where traders sneered behind conspiratorial smiles. Gypter did not mind. He made good money, making tat: pick your animal, he could fake it. Better still, his conscience stayed clean, while the tourists enjoyed a guilty, but harmless, secret thrill.

Tourists and trade would have been disagreeably impressed, had they tried to meddle with this piece of tat. Pressing a recessed switch, he recited carefully:

"In Xanadu did Kubla Khan a stately pleasure-dome decree."

"*Wrong,*" snarled a tiny voice.

He repeated the quotation.

"*One more mistake,*" gloated the voice, "*and I explode.*"

Did he have the right package? Had the voice been *quite* so venomous during rehearsals?

Again Gypter said the words.

"*Ten, nine, eight,…*"

He let it count.

"*…two, one, zero.*"

A sullen pause ensued.

The box clicked open.

Gypter breathed out.

Slotted into plastic foam were ranks of gaudy fluff and wire which a casual eye should mistake for fishing flies. Indeed, he had insisted they be usable as such, lest the box come into strange hands and, however improbably, be opened intact.

Holding the box aloft, he pressed the switch once more.

A colourful cloud whirled up into the sky. The forests were full of insects. A few extra would hardly be noticed. But every one of these should trace a course, albeit an erratic one, that returned it to the mansion within an hour.

The box now held neatly ordered naked fishhooks. It was of no further use. Tossing it into deep undergrowth, he redirected his attention to the business of enjoying his walk.

Soon after, a muffled boom sent flocks of birds surging from the trees. He watched them in mild interest, before moving on.

The dell was a lake of blue flowers. Like a pool of…

Behind eyes that had witnessed coolly the career deaths of many a foe, a struggle commenced as Gypter tried to dredge the requisite terminology from neglected poetic boglands of his mind. Fallen heaven? Nah, mushy. Scintillating cerulean? Laboured. God, this place was getting to him. *Health warning: too much scenery can soften your brain*. He gave up. It would come.

Strolling into the sunshine, he bent to pick a flower. As his fingers took hold and pulled, someone shouted at him.

"*No!*"

He froze for a second, then looked around.

She was standing, motionless, shaded by nearby trees. His snap judgment assessed her: late teens; healthy; build not suited to physical labour; damn sexy; expression suggesting he'd better not voice that last thought…

But her clothing? It was absurd. A flimsy green negligée that somehow guarded her modesty despite a frisky breeze—and nothing else. Impure thoughts simmered in his mind.

Billows of soft brown hair framed a flawless suntanned face. Precise features. Delicate nose. A mouth whose true shape was hard to gauge, being small and pursed with anger.

Green eyes were intent upon the wisp of blue in his grip.

"This?" he asked.

She nodded slowly, before sighing. "Too late."

"And if I buy into this estate?" he demanded, angered for no reason he could fathom, except finding himself on the defensive. "I'd control it all, right? Every last weed."

Her temper sparked. "Sometimes the things we suppose are ours were only hired to us by the real owners."

"Yah? And who might those be?"

"The children." Perhaps it was the movement of her head which caught another light; the green in her eyes had shifted to blue. She relaxed. "Wherever. Whosoever."

He thought to discard the flower. Of a sudden, she slipped from shade into sun and down the grassy bank, to stand before him. So gently he scarcely thought to resist, she confiscated the bloom. Twirling it beneath her nose, she smiled up at him.

"You picked it," she said. "Don't waste it."

She lodged the slender stem in the topmost teeth of his jacket's front fastener and stepped back, regarding the effect. Blue shone against military drab.

He nearly brushed it to the ground. Impulse and irritation died before his hand had moved halfway. Instead he fingered the petals.

She was studying him when he raised his eyes.

"And who are you?" he asked.

"One who lives here." A hand described an arc that implied the enclosing vista—trees, valley, sky, all.

"In the woods?"

"Of course." Laughter broke from her: innocent, without a hint of mockery. "Where else could anyone wish to live?"

"How about in a house with a roof?" There was nothing about her to suggest a life of sleeping rough. This was some damnfool joke. Lourat's people, or Lourat herself.

"That's silly. Why hide in a box?" A grimace warped her face. "A dead box, too. Ugh."

"How long have you been here?" He was careful to keep his expression neutral. If observers were hoping to laugh themselves silly, he was not about to grant them the pleasure.

She seemed puzzled. "Always. As long as I can remember. Just me. And my friends." Her sunny smile returned. "Except the animals, of course, and the plants. But animals and plants don't really count as company, even though they say nice things. I mean, that's all they ever do say. Nice things. And the rest are mostly wrapped up in their own affairs. So I suppose I live alone. Or did, until you came."

"Ah," he extemporised. Where were those cameras? Perhaps he could foil them by moving. "Shall we walk?"

Aggressively he strode along the path and up the steepening slope. She followed by his side, chattering gaily about the natural woodland life she supposedly led.

The going was wet and soft. He observed how freely she moved, skipping more than walking, each foot hardly pausing to tread the ground, as though only token contact were required, instead of the solid, squelching imprints his own boots made.

"And who are the rest?" he asked.

"People," she said.

"Like you?"

"Oh no. I'm the only dryad 'round here." She danced ahead, up to the crest.

Gypter stopped. Dryad? As in tree spirit? This was past getting absurd. If he cooperated with the gag an instant longer, they'd be right to mock him.

He joined her.

"Dryad, hah?" He studied her honest, open face. "And your friends are the Tooth Fairy, the Easter Bunny and Father Frost?"

She frowned. "No…. That is, I've never met them."

Shaking his head, he walked down the next slope. The ground levelled out into a meadow surrounded by mature oaks, beneath whose gnarled branches the path led. He ignored the girl, now frolicking amongst beds of flowers. She was singing to herself—and, for all he knew, to the flowers also.

At the first tree he halted. The sun was hot, the shade welcome. A breeze saved the place from being unpleasantly humid. By the looks of it, in wet weather the ground here became boggy. In patches the short meadow grass had been churned to mud.

Forsaking the shade, he went for a closer look. Everywhere were hoofprints. Numerous horses had rampaged to and fro, though concentrating their frenzy on two wallows at opposite ends of the clear area. By the outer fringes of each wallow, sticks had been jammed upright as if to form crude goals.

Equestrian polo? Here?

A gaudy squadron of unusually big butterflies flitted across the clearing in an undulating line and disappeared among the far trees. For a few seconds the sounds of singing seemed louder.

The girl joined him.

"For you," she declared. A bunch of flowers, gloriously varicoloured and sweetly scented, was thrust under his nose.

He inspected it dubiously. "I thought you didn't approve of people picking flowers?"

"Oh, I asked first, of course."

"Of course." Despite himself, Gypter accepted the gift. He sniffed. It did smell wonderful.

He pointed to the mud. "What happened here?"

She looked up and down the meadow.

"That would be the centaurs. Some are practising for the hoofball championship…"

She stopped when he burst out laughing.

"I don't see what's so very funny about that," she snapped. "They stand an excellent chance this year. The satyrs and fauns can barely muster a whole team between them, after last winter."

"I surrender," he said. He turned abruptly and headed back towards the parked aircycle.

The girl caught up. "Have I offended?"

"Who, you? Not in the least. Your lot have a really neat style in humour. Dryads, centaurs, satyrs. Et bloody cetera—"

With astonishing strength, she grabbed his sleeve. Fabric ripped as he lost his balance and tumbled to the ground, sending the flowers flying. All of a sudden Gypter found himself on his back, pinned beneath a babbling fury that brandished a fist and hurled words at him.

"You listen to me mister high-and-mighty barging in here as if you have a right to buy these forests and everything in them I'll have you know there are people with every bit as much right as you might fancy you own…"

In a while she lost her head of steam and ceased shouting.

Gypter looked at her.

She looked at him.

Each second of silence, in its own way, seemed fully as noisy and significant as any which had preceded it. But at last she gave a small shake of her head. Standing, she turned away.

Gypter got up, dusting off dirt and dead leaves.

"Sorry," she said. "Shouldn't have lost my manners."

"Nix," he replied. Wryly he examined the ripped sleeve. The fabric was to full Mil-Spec.

"Did I hurt you?" she asked.

"Just my pride."

She hesitated. "It's because we're worried."

"Who, about what?"

"My friends and I. About what plans you have for our home."

"You know of the deal?"

"Why not? We may shun your world. That doesn't mean we have to be ignorant of it. The people in the House tell us a little, now and then."

He fingered the ripped sleeve. "You did this so easily. I certainly couldn't. What *are* you?"

"I already said."

"Don't kid a kidder, kid."

"Dryad is near enough. It is the form I wear. The function I fulfill." Rubbing her fingertips together, she gripped the torn edges of the sleeve, massaged them for a few seconds and let go.

The join was far from perfect, yet synthetic fibres had been fused along a neat seam.

She peered up into Gypter's eyes, her head tilted sideways, grinning mischievously. "Is that sufficiently eldritch for you?"

He thrust his hands deep into his pockets for the reassuring warmth. Disoriented. Dizzy. "You realise, where I come from, people like you and your friends are thought to be imaginary."

She studied him, as though considering his sincerity.

"Be easy," she said. "We are."

"Meaning?"

"Miz Lourat knows some very clever people."

Gypter managed to find a rock to sit on. The stories about the Lourat Association and its researches had often been bizarre. Market analysts scoffed at the more orthodox speculations. What would they make of this situation?

"Just *how* clever? Genetic metamorphing? Asurgical melds? Semimentive prosthetics?"

Any of those, blue-sky babbling though they were, made a lot more sense than creatures from ancient Greek folk tales.

She laughed. "Oh, none of that. Be more direct."

Instinct warned him to duck and run. Experience demurred, goaded by curiosity. Too late, both argued. Go on. Or under.

He closed in.

"These clever people, would they do business with friends of Miz Lourat?"

"They would consider it their bounden duty."

"Even to major lawbreaking?"

She gave a slow nod. Hair flowed in across her face, hiding it. "Yes," she whispered.

Gypter brushed aside her hair. His fingers tingled.

"We must be clear on a few points."

She raised her face again. He lowered his hand.

"There are laws—good and necessary ones—which prohibit certain kinds of research. The punishments for breaking them are hideously severe. Does Lourat understand this?"

A nod.

He hesitated. "And you, are you an android?"

Gently, she nestled a hand in his. "You decide."

Her skin was soft and warm. No question, she felt the real thing. Tantalisingly so. With difficulty, he drew away.

"Think of prosthetics," she said. "Or clothing."

He raised an eyebrow.

"You're a suit of clothes?"

She laughed merrily. "Some ideas do sound funny, at first."

"Yah, right," Gypter said.

I'm talking to a Something with a weird identity problem, he thought, keeping his face straight. *Stay? Or make my excuses—and lose touch with some incredible technology…*

Easy one.

"Me," he said, "I don't care what you are. My worry is that Lourat plainly could end her money problems in a zipped minute by licensing these discoveries. If she keeps them secret, punitive seizures by the government could cost her everything. To opt for risk, instead of easy wealth and safety, she has to be crazy."

The dryad smiled sadly. "Crazy indeed. Crazy with love."

"Love of what? Look, at Lourat's age, execution and penury may hold few terrors. But her clever people, what keeps them so loyal? Doesn't she feel any responsibility?"

"As to that, you must ask Miz Lourat." She settled herself beside him on the rock. "But it could be useful to recall that love and loyalty are not always bedfellows." Shading her eyes from the sun, she admired the scenery.

Allowing me time to think, he realised.

"If I came in on this deal," he said, "would I regret it?"

She toyed with the hem of her garment.

"Would I?"

"Who can trust the future?" she said, at last. "But I don't believe *you* ever would."

"There are secrets here," he suggested.

"No outsider can be allowed to know."

Gypter thought about the mad old woman, watching her wealth drain away, increasingly fearful that her last beloved scrap of unsullied planet would fall into the clutches of despoilers. One day she, or people loyal to her, had made a discovery that reeked of money. Ironically powerless to exploit it, in her desperation she sought him out for help. Him, of all people.

"The ones in the House, they depend on you, Sir Henry."

To fix the hash they've made of things. Quite believable.

"Henry," he corrected automatically.

"As do we. Miz Lourat knows your weakness, you see."

"Which weakness?" He ran his eyes down her.

"Not that one," she said. "The streak of altruistic romance you try to hide. When you were younger, you might have sold us out for what you could make. Now that you have more money than you know what to do with, well, we were hoping you might like to have some place you could visit, now and then, when the world out there becomes too sordid to be tolerated."

Words seemed superfluous. For a while they sat together on the rock in the hot sun. Then they walked back to the aircycle. As he was about to close the door, she leaned forward to kiss him on the cheek.

"Take a cheerful thought with you," she said. "Back to your world of dead houses with roofs on them."

By the time Gypter had garaged the aircycle and regained the privacy of his apartment, the flying bugs had all found roosts. Checking only that they responded to signals and were recording, he moved to the desk in the office, where a bundle of partnership papers, tied in the customary red ribbon, waited for him.

ANDREW STEPHENSON

He scanned the text quickly by eye, then fed it through his document analyser. Within seconds the link back to his corporate noeton had returned confirmation that there were no semantic or legal traps. Appended was a demotic reduction of the tortuous legalese from which the agreement had been constructed. This he took to a window seat, where he relaxed while digesting all the proposals.

As Lourat had said, there was nothing to which he cared to object. It was a damn good deal. He would have the controlling interest for life, in exchange for a specified input of funds now and his continuing administrative expertise.

Hmm. *For life*. He leaned back to regard the ceiling, where a flourish of plaster cherubs cavorted with maidens amid a sylvan setting.

The creatures in the woods, and the secret they represented, were a whopper of an asset. Beside them, the estate was by no means nothing; but the one did somewhat overshadow the other.

And whatever Lourat had, her backyard was full of it. Them. Vibrantly alive. More than human. But hiding from humanity, reluctant to be embroiled in its doings. Okay, no need to bother these ones. But the technology underpinning them would be worth a fortune, one day soon. Instinct insisted it was so.

Control of that kind of potential could tempt a lot of folk.

Time, he decided, for a chat with Lourat.

He found her on the sunny south side of the mansion.

A broad patio was littered with fatuously elegant wrought iron garden furniture. Someone had smothered the iron in pink enamel of painful hue before infecting it with gilt measles and padding the seats with bloated mauve cushions. Candy-striped umbrellas cast shady pools, in one of which sat Lourat, swaddled by green polyester lace. The debris of a snack lay on the table beside her. Dark glasses with fluorescent orange butterfly-wing frames masked her eyes as she watched a group of stockholders and ecofreaks at play on the adjacent croquet lawn.

She raised her head to regard him as he approached.

"Henry," she murmured. "Did you have a pleasant walk?"

He sat down heavily, allowing his chair to scrape noisily. A couple of the players scowled. He ignored them, focusing on Lourat.

"Pat," he said, "I rather think you know I did."

"And the papers? Are they acceptable?"

"Very."

"But…?"

"Got to tinker with the phrasing. If I die unexpectedly, it could be, um, awkward. There's a clause in my will requiring the most exhaustive of post-mortems and formal inquiries…"

She raised a hand and smiled indulgently.

"Please, Henry. Make whatever changes you wish. In the end it will come down to trust. You trust us, we trust you."

"Yah, right," he said. His eyes strayed to the lawn, where one of the 'freaks was on her knees, gauging a shot. "And them?" He indicated the players. "Do your supporters know what you've got hidden in the woods?"

Lourat removed her sunglasses and fixed him with her bright gaze. "More to the point, Henry, do you?"

He thought about it. "Not really."

"Some lunch while we talk?" She glanced towards the clutter of used tableware. Salady stuff, mostly.

"Yah, sure. And a chilled white wine, if you have any." Gazing about the patio, he asked, "Do we send smoke signals?"

"Swanson will be along shortly."

ANDREW STEPHENSON

About thirty seconds later the footman rounded the corner of the building, a tray balanced on one hand. Gypter whistled and waved.

Swanson ignored the hail. Advancing at an unbroken pace, he reached the table and began unloading a selection of items, which together duplicated what Lourat appeared to have had.

Plus one glass of chilled white wine.

Gypter considered the wine. Swanson departed. Lourat set her glasses on her nose and delicately pushed them into place with a fingertip.

"Bit of a mind reader," said Gypter. He sipped the wine.

It was excellent.

"Dear Swanson," said Lourat, dreamily, "such a treasure."

She had not misunderstood.

"And the girl in the woods?"

"A construct."

Wood clacked against wood, out on the lawn, and players cheered. Their chattering faded into distance. On the patio was a stillness. At such moments, worlds change.

"She seemed very real."

"One of our best." Lourat laughed. "Oh, wait until you meet the others."

"Centaurs and all?"

"Whatever you want, we can build it."

"Realistic, but not truly real."

The old lady grinned. "Henry," she purred, "is that you doubting their lifelike qualities?"

He knew he was losing control. A hot flush filled his face. Too soon to blame it on the wine.

"And Swanson?" he demanded.

"Nobody could be that perfect," she giggled, settling back in her chair.

"Or that insufferable," he agreed. "How did you solve the problem of personality?"

"In Swanson's case, we inscribed an artificial one. Machine generated, apparently. The details are utterly beyond me. That sort of thing I leave to my clever people."

Instinct was yammering at him to shut up, not to ask the next question. Yet, not knowing would be…

"You are wondering," said Lourat, "about the girl."

Gypter nodded.

"Henry, I have no idea who she was. In this tragic world of ours, death happens, unremarked, all the time. My people went looking and they found her. How it works—"

"Yah, don't tell me, the details are utterly beyond you."

Gulping the rest of the wine, Gypter set the empty glass on the table and flicked it away from the edge. Instead of sliding to safety, it toppled and rolled along an arc. Calmly he watched it teeter on the brink, then smash onto the paving.

Lourat remarked, "Swanson will be most unhappy."

"Good. Having some kind of emotion should be an interesting experience for him." Gypter fought to control himself. "Gods, woman, why did you invite me into this mess?"

"As I said yesterday, you want control, we want survival."

It was a truth he could not deny. The reassurance of being the one who wields power, in a world wherein being powerless was to be miserable beyond measure, was worth more than money. To be free to choose one's future, know that another's whim would not moderate the morrow. To survive. Yes, he understood.

So why did his heart flinch from this opportunity?

The salad, he noticed, was drying out. And the players had finished their game and had slipped away.

We all lose our freshness. One by one, we all slip away.

Lourat was watching him.

"If I became part of that 'we,' what then?" he asked.

"We take care of our own," she said.

In his dreaming that night, Gypter awoke.

Someone was knocking on his door. Softly, irresistibly, the sounds summoned him from the refuge of sleep, unwilling yet obedient. Despite himself he called for entry.

The door swung wide. Swanson stood there. In one upraised hand he bore what seemed to be a coat hanger—but the garments arrayed on it were wrongly textured, horribly malformed, tucked and folded in suggestive ways.

"Sir," he said, "the mistress instructed me to bring to you this more suitable apparel."

The hand held out its burden. And a sleeve of the proffered garments came untucked. Sliding free with the rubbery grace of a deflated inner tube, it flopped and swung heavily in plain sight, so that Gypter could make out the human thumb and human fingers with which it was tipped.

He was glad to escape the place. The signed papers sat in his case like a primed bomb, awaiting validation and registration with the regulatory authorities. Concern enough, of itself. But the lingering shock of his dream hung upon him too, like a heavy cloak of grief, as if he knew he should be in mourning, without knowing what for. Unable to shake the sensation, he sat dazed in his seat, watching the ground fall away and the mansion dwindle into the scenery, aware only of the apprehension of some threat.

The cause was not the deal. Now that he knew what he must bring to market, business should proceed like any other he had shepherded to success in his time: hide connections; ensure exploitation was seen to proceed as the world expected while the real development went on covertly; set up parallel operations, slush funds and subornments; buy politicians; infiltrate pressure groups; change whichever 'freak-, church- and/or union-inspired laws required changing. Refashion prejudices. Remake a world.

The prospect should have delighted him.

As a distraction, he opened his case and began scanning bug channels. The comms monitor had been logging activity within the mansion and quickly brought him up to date.

The thought came to him, suddenly, that the bugs had been a waste of effort. From the start, Lourat had been eager to hand him the farm on a plate, simply in exchange for his help.

It was just as well. The only recorded voices were his own and those of people who had been in his presence.

He frowned, requested reanalysis. The result was confirmed: except when he was nearby, no one had spoken around the mansion since the bugs had infiltrated themselves.

But…

A channel carried voices now.

"Swanson."

"Yes, ma'am?"

"Signal our friends in the forest. Inform the contact unit that we are pleased with her performance. Advise the rest it is safe to emerge. Remind the fairies not to show off."

"At once, ma'am."

"And Swanson."

"Ma'am?"

"Have the buildings swept."

"Thoroughly, ma'am."

A closing door. And silence.

Soon after, all bugs stopped sending, nearly simultaneously. Gypter did not care. He was picking his way through a thornbush of additional unease which the brief conversation had inspired.

Someone was conning someone, above and beyond the deceits a wise person took for granted. It seemed Lourat had known of his bugs, expected him to be listening. He had been sent a message.

Why should she seek to spice a deal, already settled, with redundant talk of the creatures in the woods? She knew he knew about them.

A horrible suspicion formed in his mind.

He had been told much. He was an insider now. But much had not been told him. Precisely *what* did Lourat want him to deduce in addition, unaided?

And then another link connected itself for him, as an icy blast of fear scoured his soul.

Lourat's message to him was not in her words.

Her message was the fact of the message.

See, Henry, your bugs are working; but I only speak aloud when I must.

Patricia Amelia Lourat. She of the clear eyes amid papery skin. She of the surprising turns of agility. She who would do what she could to protect the heritage of a dead mistress...

Or of herself? Did This Year's Model transcend the mimicry of life, to the point of perpetuating it?

When had Pat Lourat died? In what secret grave did her body moulder? Beneath some forest tree, he guessed, flowers growing over her head. Where dryads wandered, singing. Where centaurs romped in sport.

A wonderland which he, Henry Gypter, ached to conceal and preserve and delight in.

When had Pat Lourat shrugged off her old body and donned the new? And when had she started to offer that kind of resurrection to others?

A gift which he, Henry Gypter, would need one day.

She knew it. Had planned for it.

He knew it too—and shuddered at his dependence on her.

Okay, he was in. The first hand had been dealt. Time to play. *Best poker face, Henry. These are sharp folks.*

It would be a game without equal. Exquisitely dangerous, virtually unwinnable, played with a deck stacked against him. At stake were life and sanity. A fortune, too, on the side. A game worthy of him.

Lourat could award extensions of life. Or withhold them. The political power which that conferred was beyond measurement. In the Coming Age, some would be content to buy; but many would seek to take and not be delicate about it. Either way, the world was going to be different, once the secret broke.

And sanity? *That* needed digesting. He had noted how Lourat presided over her obedient herd of servants and house guests, and over the beings in the woods, all of them vassals to her whims...

What servile niche had she prepared for him?

Dear Henry: such a treasure.

God, no! To endure eternity, imposed upon by her abominably wretched aesthetics. A pink and mimsy hell...

The helicopter banked. Beneath, the landscape tilted. In a forest glade, far below, Gypter fancied he saw a mob of humanoids rushing wildly in pursuit of a ball. Their legs were crooked, as if tipped with hooves. And dark, as though furry.

On their heads would be small decorative horns.

He chuckled grimly. "Give it up fellas. Word is, our team hasn't a chance. Not this year."

But, maybe eventually.

If he could restack the deck his way.

HEART OF WHITENESSE

BY HOWARD WALDROP

For John Clute: the hum of pleroma
"Doctor Faustus? —He's dead."

Down these mean cobbled lanes a man must go, methinks, especially when out before larkrise, if larks there still be within a thousand mile of this bone-breaking cold. From the Rus to Spain the world is locked in snow and ice, a sheet of blue glass. There was no summer to speak of; bread is dear, and in France we hear they are eating each other up, like the Carribals of the Western Indies.

It's bad enow I rehearse a play at the Rose, that I work away on the poem of the celebrated Hero and Leander, that life seems more like a jakes each day. Then some unseen toady comes knocking on the door and slips a note through the latchhole this early, the pounding fist matching that in my head.

I'd come up from the covers and poured myself a cup of malmsey you could have drowned a pygmy in, then dressed as best I could, and made my way out into this cold world.

Shoreditch was dismal in the best of times, and this wasn't it.

And what do I see on gaining the lane but a man making steaming water into the street-ditch from a great bull pizzle of an accouterment.

He sees me and winks.

I winks back.

His wink said I see you're interested.

My wink back says I'm usually interested but not at this instant but keep me in mind if you see me again.

He immediately smiles, then turns his picauventure beard toward the cold row of houses to his left.

Winking is the silent language full of nuance and detail: we are after all talking about the overtures to a capital offence.

I come to the shop on the note, I go in; though I've never been there before I know I can ignore the fellows working there (it is a dyer's, full of boiling vats and acrid smells and steam; at least it is *warm*) and go through a door up some rude steps, to go through another plated with strips of iron, and into the presence of a High Lord of the Realm.

He is signing something, he sees me and slides the paper under another; it is probably the names of people soon to decorate a bridge or fence.

This social interaction is, too, full of nuance; one of them is that we two pretend not to know who the other is. Sometimes *their* names are Cecil, Stansfield, Salisbury, sometimes not. Sometimes my name is Christopher, or Chris, familiar Kit, or the Poet, or plain Marlowe. We do pretend, though, we have no names, that we are the impersonal representatives of great ideas and forces, moved by large motives like the clockwork Heavens themselves.

"A certain person needs enquiring about," said the man behind the small table. "Earlier enquiries have proved…ineffectual. It has been thought best the next devolved to yourself. This person is beyond Oxford; make arrangements, go there quietly. Once in Oxford," he said, taking out of his sleeve a document with a wax seal upon it and laying it on the table, "you may open these, your instructions and knowledges; follow them to the letter. At a certain point, if you must follow them—thoroughly," he said, coming down hard upon the word, "we shall require a token of faith."

He was telling me without saying that I was to see someone, do something to change their mind, or keep them from continuing a present course. Failing that I was to bring back to London their heart, as in the old story of the evil step-queen, the huntsman, and the beautiful girl who ended up consorting with forest dwarves, eating poison, and so forth.

I nodded, which was all I was required to do.

But he had not as yet handed me the missive, which meant he was not through.

He leaned back in his chair.

"I said your name was put forth," he said, "for this endeavor. But not by me. I know you to be a godless man, a blasphemer, most probably an invert. I so hate that the business of true good government makes occasional use of such as *you*. But the awkward circumstances of this mission, shall we say, makes some of your peccadilloes absolute necessities. *Only* this would make me have any dealings with you whatsomever. There will come a reckoning one fine day."

Since he had violated the unspoken tenets of the arrangement by speaking to me personally, and, moreover, telling the plain unvarnished truth, and he knew it, I felt justified in my answer. My answer was, "As you say, Lord _____," and I used his name.

He clenched the arms of his chair, started up. Then he calmed himself, settled again. His eyes went to the other papers before him.

"I believe that is all," he said, and handed me the document.

I picked it up, turned and left.

Well, work on Hero and Leander's right out for a few days, but I betook me as fast as the icy ways would let, from my precincts in Shoreditch through the city. Normally it would mean going about over London Bridge, but as I was in a hurry I walked straight across the River directly opposite the Rose to the theatre itself in Southwark.

The River was, and had been for two months, frozen to a depth of five feet all the way to Gravesend. Small boys ran back and forth across the river. Here and there were set up booths with stiff frozen awnings; the largest concatenation of them was farther up past the town at Windsor, where Her Majesty the Queen had proclaimed a Frost Fair and set up a Royal Pavilion. A man with a bucket and axe was chopping the River for chunks of water. Others walked the ice and beat at limbs and timbers embedded in it—free firewood was free, in any weather. A thick pall of smoke hung over London town, every fire lit. A bank of heavy cloud hung farther north than that. There were tales that when the great cold had come, two months agone, flocks of birds in flight had fallen to the ground and shattered; cattle froze standing.

To make matters worse, the Plague, which had closed the theatres for three months this last, long-forgotten summer, had not gone completely away, as all hoped, and was still taking thirty a week on the bills of mortality. It would probably be back again this summer and close the Theater, the Curtain and the Rose once more. Lord

Strange's and Lord Nottingham's Men would again have to take touring the provinces beyond seven mile from London.

But as for now, cold or no, at the Rose, we put on plays each afternoon without snow in the open-air ring. At the moment we do poor old Greene's *Friar Bacon and Friar Bungay*—Greene not dead these seven months, exploded from dropsy in a flop, they sold the clothes off him and buried him in a diaper with a wreath of laurel about his head—we rehearse mine own *Massacre at Paris*, and Shaxber's *Harry Sixt*, while we play his *T. Andronicus* alternate with Thomas Kyd's *Spanish Tragedy*, of which *Andronicus* is an overheated feeble Romanish imitation.

Shaxber's also writing a longish poem, his on the celebrated Venus and Adonis, which at this rate will be done before my Hero. This man, the same age as me, bears watching. Unlike when I did at Cambridge, I take no part in the Acting; Will Shaxber is forever being messenger, third murderer, courtier; he tugs ropes when engines are needed; he counts receipts, he makes himself useful withal.

No one here this early but Will Kemp; he snores as usual on his bed of straw and ticking in the 'tiring house above and behind the stage. He sounds the bear that's eaten All the dogs on a good day at the Pit. I find some ink (almost frozen) and leave a note for John Alleyn to take over for me, pleading urgent business *down* country, to throw off the scent, and make my way, this time over the Bridge, back to Shoreditch.

Shoreditch is the place actors live, since it was close to the original theatres, and so it is the place actors die. Often enough first news you hear on a morning is "another actor dead in Shoreditch." Never East Cheap, or Spital Fields, not even Southwark itself; always

Shoreditch. At a tavern, at their lodgings, in the street itself. Turn them over; if it's not the plague, it's another actor dead from a knife, fists, drink, pox, for all that matter cannonfire or hailstones in the remembered summers.

I make arrangements; I realign myself to other stations; my sword stays in its corner, my new hat, my velvet doublet all untied, hung on their hooks. I put on round slops, a leathern tunic, I cut away my beard; in place of sword a ten-inch poignard, a pointed slouch-hat, a large sack for my back.

In an hour I am back at River-side, appearing as the third of the three P's in John Heywood's *The Four PPPP's*, a 'pothecary, ready to make my way like him, at least as far as Oxenford.

The ferrymen are all on holiday, their boats put up on timbers above the ice. Here and there people skate, run shoed on the ice, slip and fall; the gaiety seems forced, not like the fierce abandon of the early days of the Great Frost. But I have been watching on my sojourn each day to and from the Rose, and I lick my finger and stick it up (the spit freezing almost at once) to test the wind, and as I know the wind, and I know my man, I walk about halfway out on the solid Thames and wait.

As I wait, I see two figures dressed much like the two Ambassadores From Poland in my *Massacre At Paris* (that is, not very well, one of them being Kemp) saunter toward me on the dull grey ice. I know them to be a man named Frizier and one named Skeres, Gram and Nicholas I believe, both to be bought for a shilling in any trial, both doing the occasional cony-catching, gulling and sharping; both men I have seen in taverns in Shoreditch, in Deptford, along the docks, working the theatres.

There is little way they can know me, so I assume they have taken me for a mark as it slowly becomes apparent they are approaching *me*. Their opening line, on feigning recognition, will be, "Ho, sir, are you not a man from (Hereford) (Cheshire) (Luddington) known to my Cousin Jim?"

They are closer, but they say to my surprise, "Seems the man is late this day, Ingo."

"That he be, Nick."

They are waiting for the same thing I am. They take no notice of me standing but twenty feet away.

"Bedamn me if it's not the fastest thing I ever seen," says one.

"I have seen the cheater-cat of Africa," says the other, "and this man would leave it standing."

"I believe you to be right."

And far down the ice, toward where the tide would be, I spy my man just before they do. If you do not know for what you look, you will think your eyes have blemished and twitched. For what comes comes fast and eclipses the background at a prodigious rate.

I drop my pack to the ground and slowly hold up a signal-jack and wave it back and forth.

"Bedamn me," says one of the men, "but he's turning this way."

"How does he stop it?" asks the other, looking for shelter from the approaching apparition.

And with a grating and a great screech and plume of powdered ice, the thing turns to us and slows. It is a ship, long and thin, up on high thin rails like a sleigh, with a mast amidships and a jib up front, and as the thing slows (great double booms of teethed iron have fallen from the stern where a keelboard should be) the sails luff and come down, and the thing stops three feet from me, the stinging curtain of ice falling around me.

"Who flies Frobisher's flag?" came a voice from the back. Then up from the hull comes a huge man and throws a round anchor out onto the frozen Thames.

"I," I said. "A man who's seen you come by here these last weeks punctually. A man who marvels at the speed of your craft. And," I said, "an apothecary who needs must get to Oxford, as quick as he can."

The huge man was bearded and wore furs and a round hat in the Russian manner of some Arctic beast. "So you spoil my tack by showing my old Admiral's flag? Who'd you sail with, man? Drake? Hawkins? Raleigh, Sir Walter Tobacco himself? You weren't with Admiral Martin, else I'd know you, that's for sure."

"Never a one," said I. "My brother was with Hawkins when he shot the pantaloons off Don Iago off Portsmouth. My cousin, with one good eye before the Armada, and one bad one after, was with Raleigh."

"So you're no salt?"

"Not whatsoever."

"Where's your brother and cousin now?"

"They swallowed the anchor."

He laughed. "That so? Retired to land, eh? Some can take the sea, some can't. Captain Jack Cheese, at your service. Where is it you need to go, Oxford? Hop in, I'll have you there in two hours."

"Did you hear that, Gram?" asked one of the men. "Oxford in two hours!"

"There's no such way he can do no such thing!" said the other, looking at Captain Cheese.

"Is that money I hear talking, or only the crackling of the ice?" asked Captain Jack.

"Well, it's as much money as we have, what be that, Gram? Two fat shillings you don't make no Oxford in no two hours. As against?"

"I can use two shillings," said Jack Cheese.

"But what's *your* bet, man?" asked the other.

"Same as you. Two shillings. If you'll kill me for two shillings," he said, pulling at his furry breeks and revealing the butts of two pistols the size of boarding cannons, "I'd do the same for you."

The two looked back and forth, then said, "Agreed!"

"Climb in," said Captain Jack. "Stay low, hang tight. Ship's all yar, I've got a following wind and a snowstorm crossing north from the west, and we'll be up on one runner most the time. Say your prayers now; for I don't stop for nothing nor nobody, and I don't go back for dead men nor lost bones."

The clock struck ten as we clambered aboard. My pack just hit the decking when, with a whoop, Jack Cheese jerked a rope, the jib sprang up; wind from nowhere filled it, the back of the boat screamed and wobbled to and fro. He jerked the anchor off the ice, pulled up the ice-brakes and jerked the mainsail up and full.

People scattered to left and right and the iceboat leapt ahead with a dizzy shudder. I saw the backward-looking eyes of Frizier and Skeres close tight as they hung onto the gunwales with whitened hands, buffeting back and forth like skittle-balls.

And the docks and quays became one long blur to left and right; then we stood still and the land moved to either side as if it were being paid out like a thick grey and white painted rope.

I looked back. Jack Cheese had a big smile on his face. His white teeth showed bright against his red skin and the brown fur; I swear he was humming.

Past Richmond we went, and Cheese steered out farther toward the leftward bank as the stalls, awnings, booths and bright red of the Royal Pavilion appeared, flung themselves to our right and receded behind.

Skizz was the only sound; we sat still in the middle of the noise and the objects flickering on and off, small then large then small again, side to side. Ahead, above the River, over the whiteness of the landscape and the ice, the dark line of cloud grew darker, thicker, lower.

Skeres and Frizier lay like dead men, only their grips on the hull showing them to be conscious.

I leaned my head closer to Captain Cheese.

"A word of warning," I said. "Don't trust those two."

"Hell and damn, son," he smiled, "I don't trust *you*! Hold tight," he said, pulling something. True to his word, in the stillness, one side of the iceboat rose up two feet off the level, we sailed along with the sound halved, slowly dropped back down to both iron runners, level. I looked up. The mainsail was tight as a pair of Italian leggings.

"There goes Hampton. Coming up on Staines!" he called out so the two men in front could have heard him if they'd chosen to.

A skater flashed by inches away. "Damn fool lubber!" said Jack Cheese. "I got sea-road rights-of-way!" A deer paused, flailed away, fell and was gone, untouched behind us.

And then we went into a wall of whiteness that peppered and stung. The whole world dissolved away. I thought for an instant I had gone blind from the speed of our progress. Then I saw Captain Cheese still sitting a foot or two away. Skeres and Frizier had disappeared, as had the prow and the jibsail. I could see nothing but the section of boat I was in, the captain, the edge of the mainsail above. No river, no people, no landmarks, just snow and whiteness.

"How can you see?"

"Can't," said the captain.

"How do you know where we are?"

"Ded reckoning," he said. "Kick them up front, tell 'em to hang tight," he said. I did. When Skeres and Frizier opened their eyes, they almost screamed.

Then Cheese dropped the jib and the main and let the ice-brakes go. We came to a stop in the middle of the swirling snow, as in the middle of a void. Snowflakes the size of thalers came down. Then I made out a bulking shape a foot or two beyond the prow of the icerunner.

"Everybody out! Grab the hull. Lift, that's right. Usually have to do this myself. Step sharp. You two, point the prow up. That's it. Push. Push."

In the driven snow, the indistinct shape took form. Great timbers, planking, rocks, chunks of iron were before us, covered with ice. The two men out front put the prow over one of the icy gaps fifteen feet apart. Cheese and I lifted the stern, then climbed over after it. "Settle in, batten down," said the captain. Once more we swayed sickeningly, jerked, the sails filled, and we were gone.

"What was that?" I asked.

"Reading Weir," he said. "Just where the Kennet comes in on the portside. If we'd have hit that, we'd of been crushed like eggs. You can go to sleep now if you want. It's smooth sailing all the way in now."

But of course I couldn't. There seemed no movement, just the white blank ahead, behind, to each side.

"That would have been Wallingford," he said once. Then, a little farther on, "Abingdon, just there." We sailed on. There was a small pop in the canvas. "Damn," he said, "the wind may go contrary; I might have to tack." He watched the sail awhile, then settled back. "I was wrong," he said.

Then, "Hold tight!" Frizier or Skeres moaned.

He dropped the sails. We lost motion. I heard the icebrakes grab, saw a small curtain of crystalline ice mix with the snow. The moving, roiling whiteness became a still, roiling whiteness. The anchor hit the ice.

And, one after the other, even with us, the bells of the Oxford Tower struck noon.

"Thanks be to you," I said, "Captain Jack Cheese."

"And to you—what was your name?"

"John," I said. "Johnny Factotum."

He looked at me, put his finger aside his nose. "Oh, then, Mr Factotum," he said. I shook his hand.

"You've done me a great service," I said.

"And you me," said Captain Jack. "You've made me the easiest three shillings ever."

"Three!" yelled the two men still in the boat. "The bet was two shillings!"

"The bet was two, which I shall now take." The captain held out his hand. "The fare back to London is one more, for you both."

"What? What fare?" they asked.

"The bet was two hours to here. Which I have just done, from the tower bells in London to the campanile of Oxford. To do this, I had perforce to take you here in the time allotted, which—" and Jack Cheese turned once more to me and laid his finger to nose, "I have just done, therefore, *quod erat demonstrandum*," he said. "The wager being forfeit, either I shall bid you adieu, and give to you the freedom of the River and the Roads, or I shall drop you off in your own footprints on the London ice for a further shilling."

The two looked at each other, their eyes pewter plates in the driven snow.

"But…" one began to say.

"These my unconditional, unimprovable terms," said Captain Jack.

We were drawing a crowd of student clerks and *magisters*, who marvelled at the iceboat.

"Very well," sighed one of the men.

"The bet?" It was handed over. "The downward fare?" It, too.

"Hunker down in front, keep your heads down," said Jack Cheese and took out one of his mutton-leg pistols and laid it in front of him. "And no Spanish sissyhood!" he said. "For going downriver we don't stop for Reading Weir, we take it at speed!"

"No!"

"Abaft, all ye!" yelled Jack Cheese to the crowd. "I go upstream a pace; I turn; I come back down. If you don't leave the River now, don't blame me for loss of life and limb. No stopping Jack Cheese!" he said. The sails snapped up, the icebrake lifted, they blurred away into the upper Thames-Isis.

We all ran fast as we could from the centre of the ice. I stopped; so did half the crowd who'd come to my side of the river. The blur of Captain Jack Cheese, the hull and sails, and the frightened popped eyes at gunwale level zipped by.

The laughter of Jack Cheese came back to us as they flashed into the closing downriver snow and were gone.

And here I had been worried about him with two sharpers aboard. Done as well as any Gamaliel Ratsy, and no Spanish sissyhood, for sure. I doubt the two would twitch till they got back to London Docks.

The students were marvelling among themselves. It reminded me of my days at Cambridge, bare seven years gone.

But my purposes lay elsewhere. I walked away from the crowd, unnoticed; they were as soon lost to me in the blowing whiteness as I, them.

I sat under a pine by the River-side. From my pack I took a snaphance and started a small fire in the great snowing chill, using needles of the tree for a fragrant combustion; I filled my pipe, lit it and took in a great calming lungful of Sir Walter's Curse.

I was no doubt in the middle of the great university. I didn't care. I finished one pipeful, lit another, took in half that, ate some saltbeef and hard bread (the only kind to be found in London). Then I took from the apothecary pack, with its compartments and pockets filled with simples, emetics, herbs and powders, the document with the seal.

I read it over, twice. Then per instructions, added it to the fire.

I finished my pipe, knocked the dottles into the flame, and put it away.

The man's name was Johan Faustus, a German of Wittenberg. He was suspected, of course, of the usual—blasphemy, treason, subornation of the judiciary, atheism. The real charge, of course, was that he consorted with known Catholics—priests, prelates, the Pope himself. But what most worried the government was that he consorted with known Catholics *here*, in this realm. I was to find if he were involved in any plot; if suspicions were true, to put an end to his part in it. These things were in the document itself.

To this I added a few things I knew. That he was a doctor of both law and medicine, as so many are in this our country; that he had spent many years teaching at Wittenberg (not a notorious stronghold of the Popish Faith); that he was a magician, a conjuror, an alchemist, and, in the popular deluded notion of the times, supposed to have trafficked with Satan. There were many tales from the Continent— that he'd gulled, dazzled, conjured to and for emperors and kings— whether with the usual golden leaden ruses, arts of ledgerdemain, or the Tarot cards or whatnot, I knew not.

Very well, then. But as benighted superstitious men had written my instructions, I had to ask myself—what would a man dealing with the Devil be doing in part of a Catholic plot? The Devil has his own devices and traps, all suppose, some of them, I think, involving designs on the Popish Church itself. Will he use one religion 'gainst 'nother? Why don't men stop and think when they begin convolving their minds as to motive? Were they all absent the day brains were forged?

And why would an atheist deal with the Devil? The very professors tie themselves in knotlets of logic over just such questions as these.

Well then: let's apply William of Ockham's fine razor to this Gordian knot of high senselessness. I'll trot up to him and ask him if he's involved in any treacherous plotting. Being an atheist, in league with *both* the Devil *and* the Pope (and for all I know the Turk), he'll tell me right out the truth. If treasonable, I shall cut off his head; if not…should I cut off his head to be safe?

Enough forethought; time for action. I reached into the bottom of my peddlar's pack and took out two long curved blades like scimitars, so long and thin John Sincklo could have worn them Proportional, and attached them with thongs to the soles of my rude boots.

So equilibrized at the edge of the River I stood, and set out toward my destination which the letter had given me, Lotton near Cricklade, near the very source of the Thames-Isis.

And as I stood to begin my way norwestward the sun, as if in a poem by Chideock Tichbourne, showed itself for the first time in two long months through the overcast, as a blazing ball, flooding the sky, the snow and ice in a pure sheen of blinding light. I began to skate toward it, toward the Heart of Whitenesse itself.

Skiss skiss skiss the only sound from my skates, the pack swinging to and fro on my back; pure motion now, side to side, one arm folded behind me, the other out front as counterweight, into the blinded and blinding River before me. Past the mill at Lechlade, toward Kempsford, the sides of the Thames-Isis grew closer and rougher; past Kempsford to the edge of Cricklade itself, where the Roy comes in from the left just at the town, and turning then to right and north I go, up the River Churn, just larger than the Shoreditch in London itself. And a mile up and on the right, away from the stream, the outbuildings of a small town itself, and on a small hill beyond the town roofs, an old manor house.

I got off my skates, and unbound them and put them in the pack.

And now to ask leading questions of the rude common folk of the town.

I walked to the front of the manor house and stopped, and beheld a sight to make me furious.

Tied to a post in front of the place, a horse stood steaming in what must have been forty degrees below frost. Its coat was lathered, the foam beginning to freeze in clumps on its mane and legs. Steam came from its nostrils. That someone rode a horse like that and left it like that in weather like this made me burn. The animal regarded me with an unconcerned eye, without shivering.

I walked past it to the door of the manor house, where of course my man lived. The sun, once the bright white ball, was covered again, and going down besides. Dark would fall like a disgraced nobleman in a few moments.

I rang the great iron doorknocker three times, and three hollow booms echoed down an inner hallway. The door opened to reveal a hairy man, below the middle height. His beard flowed into his massive head of hair. His ears, which stuck out beyond that tangle, were thin and pointed. His smile was even, but two lower teeth stuck up from the bottom lip. His brows met in the middle to form one hairy ridge.

"That horse needs seeing to," I said.

He peered past me. "Oh, not *that* one," he said. "My master is expecting you, and cut the *merde*, he knows who you are and why you're here."

"To try to sell the Good Doctor simples and potions."

"Yeh, right," said the servant. "This way."

We walked down the hall. A brass head sitting on a shelf in a niche turned its eyes to follow me with its gaze as we passed. How very like Vergil.

We came to a closet doorway set at one side of the hallway.

"You can't just go in, though," said the servant, "without you're worthy. Inside this here room is a Sphinx. It'll ask you a question. You can't answer it, it eats you."

"What if I answer it?"

"Well, I guess you could eat *it*, if you've a mind to and she'll hold still. But mainly you can go through the next door; the Doctor's in."

"Have her blaze away," I said.

"Oh, that's a good one," said the servant. "I'll just stand behind the door here; she asks the first person she sees."

"You don't mind if I take out my knife, do you?"

"Take out a six-pounder cannon, for all the good it'll do you, you're not a wise man," he said.

I eased my knife from its sheath.

He opened the door. I expected either assassins, fright masks, jacks-in-boxes, some such. I stepped to the side, in case of mantraps or springarns. Nothing happened, nothing leapt out. I peered around the jamb.

Standing on a stone that led back into a cavern beyond was a woman to the waist, a four-footed leopard from there down; behind her back were wings. She was moulting, putting in new feathers here and there. She looked at me with the eyes of a cat, narrow vertical pupils. I dared not look away.

"What hassss," she asked, in a sibilant voice that echoed down the hall, "eleven fingers in the morning, lives in a high place at noon, and has no head at sundown?"

"The present Queen's late Mum," I said.

"Righto!" said the servant and closed the door. I heard a heavy weight thrash against it, the sound of scratching and tearing. The servant slammed his fist on the door. "Settle down, you!" he yelled. "There'll be plenty more dumb ones come this way."

He opened the door at the end of the hall, and I walked into the chamber of Doctor Faustus.

The room is dark but warm. A fire glows in the hearth, the walls are lined with books. There are dark marks on the high ceiling, done in other paint.

Doctor John Faust sits on a high stool before a reading stand; a lamp hangs above. I see another brass head is watching me from the wall.

"Ahem," says the servant.

"Oh?" says Faustus, looking up. "I thought you'd be alone, Wagner." He looks at me. "The others they sent weren't very bright. They barely got inside the house."

"I can imagine," I say. "Your lady's costume needs mending. The feathers aren't sewn in with double-loop stitches."

He laughs. "I am Doctor Johan Faustus."

"And I am—" I say, thinking of names.

"Please drop the mumming," he says. "I've read your *Tamerlane*— both parts."

I look around. "Can we be honest?" I ask.

"Only one of us," he says.

"I have been sent here—"

"Probably to find out to whom I owe *my* allegiance. And its treasonableness. And not being able to tell whether I'm lying, to kill me; better safe than sorry. Did you enjoy your ride on the ice?"

There is no way he could have known. I was not followed. Perhaps he is inducing; if he knows who I am, and that I was on London this morning, only one method could have gotten me here so fast. But no one else who saw—

I stopped. *This* is the way fear starts.

"Very much," I say.

"Your masters want to know if I plot for the Pope—excuse me, the Bishop of Rome. No. Or the Spaniard. No. Nor French, Jews, Turks, no. I do not plot even for myself. Now you can leave."

"And I am to take your word?" I ask.

"I'm taking yours."

"Easily enough done," I say. Wagner the servant has left the room. Faustus is very confident of himself.

"You haven't asked me if I serve the Devil," he says.

"No one serves the Devil," I say. "There is no Devil."

Faustus looks at me. "So they have finally stooped so high as to send an atheist. Then I shall have to deal with you on the same high level." He bows to me.

I bowed back.

"If you are a true atheist, and I convince you there's magic, will you take my word and go away?"

"All magic is mumbo-jumbo, sleight-of-hand, mists, leg-erdemain," I say.

"Oh, I think not," says Faustus.

"Blaze away," I say. "Convenient Wagner has gone. Next he'll no doubt appear as some smoke, a voice from a horn, a hand."

"Oh, Mr Marlowe," says Faustus. "What I serve is knowledge. I want it all. Knowledge is magic; other knowledge *leads* to magic. Where others draw back, I begin. I ask questions of Catholics, of Jews, of Spaniards, of Turks, if they have wisdom I seek. We'll find if you're a true atheist, a truly logical man. Look down."

I do. I am standing in a five-pointed star surrounded by a circle, written over with nonsense and names in Greek and Latin. Faustus steps off his stool. Onto another drawing on the floor. The room grows dark, then brighter, and much warmer as he waves his arms around like a conjuror before the weasel comes out the glove. Good trick, that.

"I tell you this as a rational man," he says. "Stay in the pentagram. Do not step out."

I felt hot breathing on the back of my neck that moved my hair.

"Do not look around," says Faustus, his voice calm and reasoned. "If you look around, you will scream. If you scream, you will jump. If you jump, you will leave the pentagram. If you leave it, the thing behind you will bite off your head; the Sphinx out yonder was but a dim stencil of what stands behind you. So do not look, no matter how much you want to."

"No," I said. "You've got it wrong. I won't look around, not because I am afraid I'll jump, but because the act of *looking* will be to admit you've touched a superstitious adytum of my brain, one left over from the savage state. I *look*, I am lost, no matter what follows."

Faust regards me anew.

"Besides," I said, "what is back there—" here whoever it was must have leaned even closer and blew hot breath down on me, though as I remember, Wagner was shorter, someone else then…"is another of your assistants. If they are going to kill me, they should have done it by now. On with your show. I am your attentive audience. Do you parade the wonders of past ages before me? Isis and Osiris and so forth? What of the past? Was Julius Caesar a redhead, as I have heard? How about Beauty? The Sphinx woman should have been able to change costumes by now?"

"You Cambridge men are always big on Homer. How about Helen of Troy?" asks Faustus.

"Is this the face that launched a thousand ships, etc.?" I ask. "I think not. Convince me, Faustus. Do your shilly-shally."

"You asked for it," he says. I expected the knife to go through my back. Whoever was behind me was breathing slowly, slowly.

Faustus waved his arms, his lips moved. He threw his arms downward. I expected smoke, sparkles, explosion. There is none.

It is fourteen feet tall. It has a head made of rocks and stones. Its body is brass; one leg of lead, the other of tin. I know this because the room was bright from the roaring fires that crackled with flame from each foot. This was more like it.

"Speak, spirit!" said Faustus.

"*Hissssk. Snarrrz. Skazzz,*" it said, or words to that effect.

And then it turned into the Queen, and the Queen turned into the King of Scotland. I don't mean someone who looked like him, I mean him. He shifted form and shape before me. He turns, his hair is longer, his nose thinner, his moustache flows. He changes to another version of himself, and his head jumped off bloodily to the floor. He turns into a huge sour-faced man, then back to someone who looks vaguely like the King of Scotland, then another; then a man and woman joined at the hip, another king, a woman, three fat Germans, a thin one, a small woman, a fat bearded man, a thin guy with a beard, a blip of light, another bearded man, a woman, a tall thin man, his son—

This was very good indeed. Would we had him at the Rose.

"Tell him of what lies before, Spirit."

"Tell him," I said to Faustus, "to tell me of plots."

"PLOTS!" the thing roared. "You want the truth?" It was back fourteen feet tall and afire, stooping under the ceiling. "You live by a government. Governments NEED plots! Else people ask why they die? Where's the bread? Human. Hu-man! You are the ones in torment! We here are FREE!

"PLOTS! BEware ESSEX!" Essex? The Queen's true right arm? Her lover? "BEWARE Guido and his dark SHINING lantern! BEWARE the House in the RYE-fields! Beware the papers in the TUB OF flour! Beware pillars! BEWARE POSTS! BeWARE the Dutch, the FRENCH, the colonists in VIRGINIA!" Virginia? They're lost? "BEWARE RUSSia and the zuLU and the DUTCHAFricans! Beware EVERYTHING! BEWARE EVERYboDY! AIIIiiiiiii!!!"

It disappears. Faustus slumps to the floor, sweating and pale.

The light comes back to normal.

"He'll be like that a few minutes," says Wagner, coming in the door with a jug of wine and three glasses. "He said malmsey's your favourite. Drink?"

We shook hands at the doorway early next morning.

"I was impressed," I said. "All that foofaraw just for me."

"If they're sending atheists, I had better get out of this country. No one will be safe."

"Goodbye," I said, putting the box in my pack. The door closed. I walked out past the hitching post. Tied to it with a leather strap was a carpenter's sawhorse. Strapped about the middle of it, hanging under it, was a huge stoppered glass bottle filled with hay. *How droll of Wagner,* I thought.

I went to the river, put on my skates, and headed back out the Churn to the Thames-Isis, back to London, uneventfully, one hand behind me, the other counterweight, the pack swinging, my skates thin and sharp.

Skizz skizz skizz.

When I got back to my lodgings, there was a note for me in the locked room. I took the token of proof with me, and went by back ways and devious alleys to an address. There waiting was *another* high lord of the realm. He saw the box in my hand, nodded. He took the corner of my sleeve, pulled me to follow him. We went through several buildings, downward, through a long tunnel, turning, turning, and came to a roomful of guards beyond a door. Then we went upstairs, passing a few clerks and other stairwells that led down, from whence came screams. Too late to stop now.

"Someone wants to hear your report besides me," said the high lord. We waited outside a room from which came the sounds of high, indistinct conversation. The door opened; a man I recognised as the royal architect came out, holding a roll of drawings under his arm, his face reddened. "What a dump!" said a loud woman's voice from the room beyond.

"What a dump! What a dump!" came a high-pitched voice over hers.

I imagined a parrot of the red Amazonian kind.

"Shut up, you!" said the woman's voice.

"What a dump! What a dump!"

"Be sure to make a leg, man," said the high lord behind me, and urged me into the room.

There she was, Gloriana herself. From the waist down it looked as if she'd been swallowed by some huge spangled velvet clam while stealing from it the pearls that adorned her torso, arms, neck and hair.

"Your Majesty," I said, dropping to my knees.

The lord bowed behind me.

"What a dump!" said the other voice. I looked over. On a high sideboard, the royal dwarf, whose name I believed to be Monarcho, was dressed as a baby in a diaper and a bonnet, his legs dangling over the sides, four feet from the floor.

"Well?" asked the Queen. "(You look horrible without your face hair.) Well?"

I nodded toward the box under my arm.

"Oh, give that to someone else; I don't want to see those things." She turned her head away, then back, becoming the Queen again.

"Were we right?"

I looked her in the eyes, below her shaven brow and the painted-in browline, at the red wig, the pearls, the sparkling clamshell of a gown.

"His last words, Majesty," I said, "were of the Bishop of Rome, and of your late cousin."

"I knew it," she said. "I knew it!"

"I knew it!" yelled Monarcho.

The Queen threw a mirror at him. He jumped down with a thud and waddled off to torment the lapdog.

"You have been of great service," she said to me. "Reward him, my lord, but not overmuch. (Don't ever appear again in my presence without at least a moustache.)"

I made the knee again.

"Leave," she said to me. "You. Stay," she said to the lord. I backed out. The door swung. "Builders!" she was yelling. "What a dump!"

"What a dum—" said Monarcho, and the door closed with a thud.

So now it is another wet summer, in May, and I am lodging in Deptford, awaiting the pleasure of the Privy Council to question me.

At first I was sure it dealt with the business of this winter last, as rumour had come back to me that Faustus had been seen alive in France. If *I* had heard, other keener ears had heard a week before.

But no! The reason they sent the bailiff for me, while I was staying at Walsingham's place in Kent, was because of that noddy-costard Kyd.

For he and his friends had published a scurrilous pamphlet a month ago. Warrants had been sworn; searches made, and in Kyd's place they found some of my writings done, while we were both usually drunk, when we roomed together three years ago cobbling together old plays. I had, in some of them, been forthright and indiscreet. Kyd even more so.

So they took him downstairs, and just showed him the tongue-tongs, and he began to peach on his 104-years-old great naunt.

Of course, he'd said *all* the writings were mine.

And now I'm having to stay in Deptford (since I can get away to Kent if ever they are through with me) and await, every morning, and the last ten mornings, the vagaries of the Privy Council. And somewhat late of each May evening, a bailiff comes out, says, "You still here?" and "They're gone; be back here in the morning."

But not *this* morning. I come in at seven o' the clock, and the bailiff says, "They specifically and especially said they'd not get to you today, be back tomorrow." I thanked him.

I walked out. A day (and a night) of freedom awaited me.

And who do I spy coming at me but my companions in the adventure of the iceboat, Nick Skeres and Ingram Frizier, along with another real piece of work I know of from the theatres (people often reach for their purses and shake hands with him) named Robert Poley.

"What ho, Chris!" he says, "how's the playhouse dodge?"

"As right as rain till the Plague comes back," I say.

I watch, but neither Skeres nor Frizier seem to recognise me; I am dressed as a gentleman again; my beard and moustache new-waxed, my hat a perfect comet of colour and dash.

"Well, we're heading for Mrs Bull's place," says Poley. "She owes us each a drink from the cards last night; it is our good fortune, and business has been good," he says, holding up parts of three wallets. "How's about we stands you a few?"

"Thanks be," I say, "but I am at liberty for the first time in days, and needs be back hot on a poem, now that Shaxber's *Venus and Adonis* is printed."

"Well, then," says Poley, "one quick drink to fire the Muse?"

And then I see that Skeres is winking at me, but not one of the winks I know. Perhaps his eye is watering. Perhaps he is crying for the Frenchmen who we hear are once again eating each other up like cannibals. Perhaps not.

Oh well, I think, what can a few drinks with a bunch of convivial invert dizzards such as myself harm me? I have been threatened with the Privy Council; I walk away untouched and unfettered.

"Right!" I say, and we head off toward Deptford and Mrs Bull's, though I keep a tight hold on my purse. "A drink could be just what the doctors ordered."

A DAY WITHOUT DAD

BY IAN WATSON

I asked Miranda at breakfast, "Will you do me a favour, darling?"

Beloved daughter looked dubious, which surprised me.

"Will you look after your grandad today?"

Miranda gazed into her bowl of oats and dried banana bits. Her flaxen hair hung around her face, hiding her expression from me.

"Can I keep him asleep, Mum? There's a French test this morning—and…there are swimming heats in the afternoon."

"Grandad can help you with the French."

"Don't be silly—that's cheating. Black mark if they find out."

When she said *cheating*, involuntarily I glanced at Paul, but just then he was looking at his watch.

When had beloved husband last made proper love to me? Not for three years, since Dad was installed as my guest. 'Are you sure he's asleep, Cath?' 'Of course I'm sure.' 'You might lose control…' In my spasm of pleasure Paul thought that Dad might surface as an uninvited spectator of his performance. I could hardly ask Miranda if she would host Dad for an hour at bedtime so that her parents could enjoy some unspecified spontaneous privacy! How embarrassing, how inhibiting.

Increasingly I suspected that in the past year or so Paul may have enjoyed a little side-dish, as it were—which he would no doubt justify to himself by some rationalization about his male urge demanding to be satisfied; as if I had become some sort of hospice nun without appetites and frustrations. Probably an occasional girl from the Rough, picked up while he was taking a Jag to some customer. A girl who would be glad of a modest gift of cash.

I wasn't about to rock the boat of our marriage. Paul was sensible. So was I. Full-blown affairs, divorces, were ruinous. These days financial considerations dominated most people's lives. Ours, certainly. Keeping up payments on house and health and winter heat and insurances and service contracts and all else. Investing for Miranda's future. Oh, let her become rich through her talent for design—surely she *was* showing flair!

Although Paul and I were only in our mid- to late-thirties, we invested obsessively for our old age so that we did not ever burden Miranda, as Dad burdened me.

"I'd gladly help out if I could," Paul murmured.

He couldn't, as he knew full well. Guesting only worked with genetically close relatives. All to do with the brain patterns.

"I could get some practice in," he joked feebly.

Not much need of that! Paul's own Mum and Dad weren't even sixty. Betty and Jack were both hale and hearty. Anyway, Paul tagged his sister Eileen as a soft touch if it ever came to hosting Betty and Jack. Both at once. How could the aging couple be separated after a lifetime together? Eileen would have to take both parents on board.

"You know I don't normally mind, Mum," Miranda said. "It's just…well, with the swimming this afternoon."

Miranda didn't wish to be in changing rooms with a seventy-odd-year-old man inside her head. Like a voyeur. If her friends knew, they would be furious, never mind that she swore she was keeping her guest suppressed. She would need to pretend all day long that she was on her own, which would be a strain, and a bit alienating for Dad too.

Oh why hadn't she mentioned those swimming heats until now? I'd been counting on her.

Answer: I didn't pay enough attention to her swimming. I wanted Miranda to concentrate on art, where she showed such budding talent. But Miranda nursed dreams of being a champion swimmer, which wouldn't bring her very much long-term money, only some transitory glory. Maybe art would be a false trail too, yet at least art might be a route to something special. Or art might be an awful blind alley—which was why she strove at swimming, imagining medals and sponsorships and product endorsements. Miranda didn't show much skill at economics or science or computer studies.

She was usually willing to give me respites from day-in-day-out-Dad. Not that Dad was intrusively present all the time, but still the sense of him was always with me, beneath the surface if not above.

What else could I have done other than accept the responsibility of having Dad in my head when he became unable to look after himself? The cost of putting him in a nursing home would have hamstrung Paul and me.

Could have been worse, I suppose. Mum might have lived long enough for guesting to be developed. Then both of them would have been sharing my brain.

So what should my own daughter do but help me out now and then?

"You can't keep a guest stifled all day long," I reminded her. "He'd become, well…"

"Stir-crazy," Paul said unhelpfully. "The isolation from everyday input. He needs to have a good six hours' experience a day."

So Paul was suddenly the expert, who had no guest and would likely never have one?

Monday to Saturday, Dad's six hours per day—or longer, ideally!—should obviously occur while I was at home, tele-selling to raise extra income, and while Paul was twenty miles away in the neighbouring Smooth at the Jaguar showroom, smiling, smiling at potential customers for those big sleek lux cars, secretly detesting his clients for their affluence; but he was a superb salesman. After a day of feigning and fawning, the last thing he wanted was to share the evening with my Dad.

We were contented, I suppose. We were surviving—and Miranda was our treasure, quite as much as our painfully accumulating investments. Just, we no longer felt at all young. Maybe this was true of the majority of people like us. Having Dad in my head didn't help.

"I'll do it tomorrow," promised Miranda.

I would have to postpone a certain matter. And count my blessings—mainly the blessing of a daughter who was at least willing to share my duty.

I was determined that when I grew old and infirm I would never impose myself upon Miranda, nor let Paul do so, either. That day was a long way off. If our investments proved inadequate, maybe Miranda might be wealthy enough to pay for our dotage.

"Tomorrow will be fine, darling."

Paul shivered. In his salesman's suit and striped shirt and Jaguar-crest tie, he was feeling chilly.

He didn't ask what I'd hoped to be doing today, and now instead would do tomorrow, unencumbered of Dad. If asked, I had intended to say that after lunch I would bundle myself up and walk to the park, to the hothouses, where entry was still free. I'd be losing potential sales, but I ached to see orchids in bloom, and be warm, and on my own. Yet if I had Dad in my head I would feel obliged to share the beauty with him, since sitting in on tele-sales could hardly be very exciting for him as his usual recreation.

I wasn't intending to go to the park at all.

The car showroom provided a sort of hothouse for Paul every day, though only polished metal was on display. For the sake of the customers the showroom needed to be kept considerably warmer during the winter than our own little terraced house.

It was time for Paul to put on his overcoat and rush to catch the bus. Likewise, in another quarter of an hour, Miranda. Dad was stirring. He liked to see his granddaughter off to school. Paul, he could miss. Paul could miss him.

My watch had a time-tally function so that I could be sure Dad enjoyed at least six hours of liberty. If I went to the toilet or when I was getting washed I would of course suppress him. The temptation, then, was to leave him dormant for longer than need be.

As I settled in front of my as-yet-blank screen, with Dad alert inside me, I announced, "Miranda has swimming heats this afternoon."

Oh I'd love to see those.

"I shan't see them, either."

I heard him as a voice inside me. For him to hear me, I needed to speak aloud. Guests didn't have access to your private thoughts, only to what you saw and said and did. Some hosts must be having a much harder time than me, if the beloved parent was cranky or overbearing. With a fractious parent inside, one might almost feel schizophrenic. Quite frequently I found myself telling myself a highly factual account of who Cath is, and of what's-what-in-the-world, as a way of affirming my own identity. I'd never yet felt the need to consult a guesting counsellor, even if the service is free.

At least no guests are downright senile, since a senile mind can't make the transfer.

She wouldn't really want you there, would she? Not because you'd put her off her stroke. But swimming's her way of being herself. Launching out.

"I don't know, Dad, if that's a cliché or if it's wisdom."

Actually, Cath, it was meant as a joke.

Poor old Dad. He did try.

Another day of insurance, eh Cath?

"What else, Dad?"

I'm becoming quite an expert in my old age.

If only Dad had been more of an expert in managing his own affairs in years gone by! I would never say this aloud; and I trusted Paul never to do so either in Dad's hearing.

Dad had been a metal sculptor, and quite well regarded in his day. The hot tang of the workshop always lingered in my memory from my childhood, a metallic taste as much as a smell. Evidently Miranda inherited her artistic gifts from Dad, these skipping a generation. During his career Dad had picked up enough commissions to make an adequate living, though never enough for any nest-egg. Back then, most people didn't realize how single-mindedly they must try to amass capital to pay for future care. And then the arthritis crept up on Dad. After Mum died, he ended up living in a single rented room, which he always proudly insisted was adequate for his diminished needs— until the arthritis worsened.

Dad: poor, and old.

His metal-work was expensive to carry out, so he and Mum had put off having their only child until they were a bit long in the tooth, unlike Paul's parents.

Penny for your thoughts?

No, Dad, no.

Don't worry about Miranda. She'll be fine.

In the swimming pool at school—or in the waters of the future, infested by the piranhas of finance?

Finally I switched the screen on.

But first…

"Dad, I need to take a leak—"

Which was true. I nudged the time-tally. With an inner impulse which was now second nature, I pushed Dad down to muse in his own memories, disconnected from what I was doing.

Chilly in the toilet. Not a room to linger in, despite the Breughel posters brightening the walls. The posters were creatively stimulating for Miranda, but a salutary warning for me and Paul. I could identify with those medieval peasants. Lots of clothes, and tight circumstances, hunger and cold and disease hovering not far away if anything went wrong.

I'd delayed long enough.

Before inviting Dad back, I called a certain number to apologize to Mrs Appleby, as she called herself: a cheerful, rosy name.

So sorry, Mrs Appleby. Can't make it this afternoon. A family problem. Please, will you find someone else? But please, I do want to do it tomorrow.

Fair enough for Mrs Appleby to point out with a sniff of disapproval that today was to have been my first…*engagement*, as she phrased it.

Neither of us were on-screen, although we'd inspected each other visually when we made the arrangement four days earlier; and I'd squirted her a swimming-costume photo of myself, taken a couple of years ago. That was when Paul and Miranda and I had enjoyed a discount weekend break at the West Midlands Tropicdome. Miranda revelled in the simulated ocean surf. I delighted in the jungle garden.

Paul was happy gambling with tokens, fantasy value only. The dome was a family venue. Mrs Appleby, whose tactfully phrased advertisement I'd discovered on the Web, had seen that I was very much the same trim person as in the photo. Still had my looks. She'd assured me that many married women were in her data-bank.

A pause, while Mrs Appleby accessed her timetable.

"Very well, dear. Tomorrow afternoon. Same place: Meridian Hotel. Room 323. Got it? Two o'clock sharp. This one's a German businessman. Late thirties. Make sure you count your cash. My agency fee's already included in the room hire, you'll recall."

"Understood, Mrs Appleby."

"Enjoy yourself, dear."

Would I? An act of neutral lust with a perfect stranger…

And a German too. Probably he'd be very polite and efficient. Being only in his thirties, he shouldn't have acquired a gut. He'd be tanned and athletic—superior to whichever Brit I was passing up. Silver linings!

I keyed for Insure. Then I nudged the time-tally and opened myself once again to Dad, and scrolled the work log.

The name Viking Industries took my fancy. I was about to call up its business profile when the name went red. One of my tele-colleagues somewhere else in the country had got there first.

Four fruitless calls to companies, and I was scanning the profile of a fifth prospect called KhanKorp. Newly registered, importers of spices. No doubt KhanKorp made its own insurance arrangements within the Asian community. Those arrangements might be lax, and fail to comply with all the increasingly elaborate legal requirements.

Bruce and the spider, Cath…

Yes, yes, Dad. When Robert the Bruce was hiding in his famous cave, he wrecked a spider's web again and again, and each time the spider rebuilt it. Moral: persevere.

In fact, I only needed to persuade half-a-dozen companies a week to accept a full free on-site survey to cover the leasing of the hardware and software from Omega Insurance, the last word in industrial protection. If one of those surveys scored a contract, I was modestly in pocket. This all still took hours and hours of time.

Spices! Cinnamon and cardamom and fenugreek...

Dad had loved curries. Gone were the days when he could taste anything at all, except maybe in memory.

So call KhanKorp up. Request a window for face-to-face. Smile, smile. Recite the spiel.

Miraculously competitive rates, fully secured reinsurance, the very latest in pollution and radon detectors and hazard sensors as per the most recent Euro regulations. Avoid crippling fines.

At all costs be friendly. Be careful of the implication of blackmail. ("Ask yourself, sir, what if the Safety Inspectorate should pay a random visit next week?")

At least I didn't need to pump flesh, as Paul must do before showing off the armour plating of Jags and the pollen and particle filters and the in-car voice-addressable terminal and all else.

KhanKorp was another wash-out. The Khan who deal with me was quite abrupt. Some Chinese name—Chung Hong, or whatever—handled all such matters. A Triad company, perhaps.

Just then, *ring-ring*. On-screen flashed a phone icon—to be replaced, as soon as I accepted, by a young woman's head and shoulders; and my own mini screen-top camera would now be showing me to her, in my black business dress which I wore over thermal underclothes.

She was quite an item. A freckled red-head. Gold lam jumpsuit. Huge hoop earrings. Letters pulsed in a spidery sunburst logo behind her.

"Catherine Neville? It's Denise Stuart at TV-NET. *Excusez* the interruption, Cath. I'm interviewing professional *femmes* who have an elderly guest…"

She would have found my name in the public register of guests and hosts.

I would be paid an adequate disturbance fee. A simple contract replaced Denise's image and expanded. I scrolled through quickly—and noticed that I was agreeing to Dad being interviewed as well as me. The human interest of this show might not merely be humanitarian—the experience of women who had accepted a mother or father, for the benefit of those contemplating such a step—but tensions, frustrations, regrets, even conflict.

The fee's worth having.

"Hmm," I mused.

I shan't say anything embarrassing. If you don't like something I say, Cath, just don't repeat it.

That could look awkward. Denise asking a leading question, and me fluffing the lines.

This could lead on to you becoming a counselling consultant or whatnot.

Could it? Dad was ever-hopeful.

You'd be good at it.

I was enough of a "professional *femme*" to have use of an up-to-date screen, courtesy of Omega. *Really* successful working women ought to be able to afford a nursing home, unless love and affection prevented them from abandoning their parent to some gerry-barracks run by a fat insurance company. Denise Stuart must be zeroing in upon struggling pro-*femmes*.

With mild misgivings, I assented. It was a change from phoning businesses with veiled threats.

"Have you heard, Cath, the Japanese announce they'll be able to store old folks' minds in terabyte computers in another two or three years?"

Was this true?

She was trying to catch me off balance. To provoke an exclamation of *thank God for that.*

"I'll believe it when I see it," was my reply. "If we could load a mind into any old brain, meat or machine, we wouldn't need to rely on close relatives."

"Would you miss being with your daughter, George?"

I was a metal-sculptor. What a joke to end up as part of a machine myself.

"Dad says, I was a metal-sculptor…," I repeated, et cetera.

"Do you ever feel *un peu* suffocated in there, George?"

It could get a bit crowded for my granddaughter when Cath grows old if Miranda has to look after her Mum and Dad and me as well.

I paraphrased.

Denise grinned. "*Bien!* Now you've raised a point there, George. We rush into new technology, don't we just? A senior citizen in care is going to die sooner or later—unless machines keep him ticking over till he reaches the statutory hundred years of age. *Ça coute cher!* That really costs! And as yet there isn't any such statutory limit on elderly guests—unless of course the host applies for evacuation for good reason—"

Evacuation! Mental abortion. Was she trying to scare my Dad?

"—whereupon there's nowhere else to put your mind if you're still under a hundred, George, since your body was already cremated, *n'est-ce pas*? The way I read the runes, evacuation's going to become mandatory when guests reach their century. Do you think your Cath will try to do a runner to some Caribbean island, *par example*, to keep you going? I mean, given the experience to date of sharing her head with you."

A Caribbean island was such a fantasy.

You're trying to cause trouble, Miz Stuart.

"You're trying to cause trouble, Denise."

"Is that what your Dad says? *Vraiment?*"

"You better believe it."

Denise was unruffled. "So how about the wisdom-of-the-ancestors stuff, eh Cath? Old folks used to be revered for all their accumulated know-how. I hear that even the Neanderthals kept their old folks on the go, chewing their food for them when things got tough. But, *il faut demander*, in a world of ever-accelerating change, what use is old wisdom? How much does your Dad contribute daily to your business activities? Does he impede you? What is your business exactly?"

She asks too many questions. Steamrollering, and showing off.

I saw the ideal conclusion to this interview, without voiding my fee.

"Since you ask *exactly*, Denise, I represent Omega Insurance, popularly known as the last word in industrial protection. For instance, I'll call you to ask if you know the latest Euro ruling on radiation emissions from electronic equipment such as must litter your studio—"

"*Merci* much, Cath! *Au revoir.*"

Denise would trim the interview at *You'd Better Believe It.* Ending on a note of truculence. She would pose the question: are hosts ever fully truthful?

Well-handled.

"You helped, Dad."

Some aging relatives committed suicide rather than imposing on their children. Doctors always helped out with a painless dose these days.

To be deprived of your own body! To be a guest on sufferance in the body of your own child who had grown up! To miss out on so much of real life, even if guests could stroll, dreamlike, down memory lane during the hours of the day when they weren't summoned to see and hear whatever was ongoing. Had Dad chosen the brave option or the cowardly option? It was hard to be sure.

As the hours passed I persuaded two businesses to accept safety surveys. One was a small manufacturer of anti-vandal paint of their own patent, which had exotic additives. Potentially hazardous.

When Miranda came home from school she was flushed as much with satisfaction as from the hard frost outside. As expected, she had won her heats easily. Dad was ever so pleased.

"I'll be looking after you all day tomorrow, Grandad," she told me cheerfully, and him within. She hadn't forgotten. "All day at school! That'll make a change, won't it?"

You didn't tell me this, Cath!

"I didn't want to disappoint."

"Oh Mum, I promised."

We'd long since got used to such conversations involving an unheard voice.

Sweater-clad, Miranda quickly got on with homework, using my screen in the work-nook to help with the maths which she less than loved. Maths might have seemed to loathe her too, were it not for the interactive program she could access at a modest cost, conducted by ever-patient, ever-friendly 'Uncle Albert.'

When Paul got home, he was fascinated to hear about the interview with Denise Stuart. It wasn't till after our meal of soy and veg, nicely spiced, that he confided modestly how he hoped to clinch a deal with a Chinese client for a top-of-the-range XJ5000, pending a final test drive through the Rough the next morning. The sale would mean several hundred in commission.

"Not a motorway sprint, but Rough driving—"

"He's probably a Triad boss," I joked. "Needs to deliver *stuff* to dodgy destinations." Miranda ought to know about such aspects of life, at least in the abstract.

"Probably is," Paul agreed. "Sam Henson says he heard how a drugs boss guested an undercover Chinese taxman into the head of a monkey a couple of months ago."

Miranda shuddered to hear this. "That's impossible, isn't it?"

"Totally," I assured her. "Your Dad's boss is a bit of a racist."

Our daughter grinned. "Is that why he sells fast cars?"

Miranda was streetwise enough—or so she imagined herself. Of course she was sheltered by living in the Smooth. No drugs or gangs or vice at her school. At least we hoped not.

"If that *could* become possible," said Miranda, "you know, I think I'd quite like to share, say, a dolphin's day—"

For the swimming, oh yes.

"—just so long as I could pop back into my own body afterwards. Like Grandad will pop back into yours, Mum." And out gushed: "Are you going somewhere special tomorrow?"

"Just to the park. To the orchid-house. To be on my own for a while."

"You deserve it," Paul said.

Miranda might worry that something could happen to me—a car skidding, say, and Mum dying. Then she would be saddled with Grandad for the rest of her life. Unless, of course, Grandad nobly insisted on evacuation, or Paul petitioned. That would be traumatic for Miranda.

I usually knew when she had something special to tell me, and finally she got round to it.

"Mum, there's a girl in the first grade called Jenny O'Brien. Her mother has had a new baby, and it's sick with leukemia. It's going to die. She's applying to guest the baby."

"The girl is?" asked Paul, incredulous.

"No, Dad, her mother is! Mrs O'Brien wants to rear the baby inside her head. Teach it. Let it have a chance. She's a Catholic, you see."

"Good God," said Paul.

I was stunned as well.

To host an unformed mind… What sort of mind would that be? Full of infantile appetites, few of which could be catered for. Would it be able to learn to see through its mother's eyes, or learn to talk?

Like a latter-day Helen Keller. At least Helen Keller had a body of her own, even if she was born blind, deaf, and dumb.

"It's meant to be a secret, Mum, but Jenny O'Brien's very upset. I was thinking I could introduce her to Grandad tomorrow. Show her that her mother won't be totally occupied with the baby."

This was so thoughtful that I almost felt ashamed of the purpose for which I was asking Miranda to look after Dad. You might say that Miranda could afford to be considerate, living here in the Smooth. On the other hand, she could have grown up snooty and selfish, with false expectations. No doubt the chill of our house in winter and the stifling heat in summer curbed any affectations and made her realistic.

"Your Denise Stuart would be interested to hear about this," Paul hinted.

"Oh no, Dad, it's private!"

"Mrs O'Brien might *need* publicity if she's to have any hope of persuading the guesting office. A campaign in her support."

"If that's so," I said, "she'd probably rather start any campaign *herself*."

Paul persevered. "This hasn't happened before, has it? There would have been publicity. It sounds like a fascinating experiment. Maybe Denise Stuart might know."

"Maybe Sam Henson might know!" I hoped I didn't sound brusque. I felt protective of Miranda's confidence in us. I didn't wish her to be at all upset this evening, nor tomorrow morning either.

Miranda sought to change the subject. "That was a lovely meal, Mum—"

"Was Uncle Albert helpful?"

She grimaced comically. "Just a bit."

When Paul and I went to bed early to keep warm, he kissed me, then he quickly turned over, lying flat upon his belly, as if otherwise I might make some physical demand upon him. My idea of an ambiguous reward for his wooing of the Triad boss.

Probably I would be deeply disappointed tomorrow, and disgusted with myself. My German paramour mightn't be courteous at all. How could he have any idea what significance his pawings and thrustings would have for me, in my imagination?

Ah, but to do something which was utterly impossible while Dad was in me! Something from which he inhibited me, as surely as he inhibited Paul!

✦

Before breakfast next morning Miranda and I made the exchange.

We sat face to face, our knees interlocked. Both of us steadied the 'binoculars,' as the transfer apparatus—domestic variety—inevitably was nicknamed.

It was a spin-off from the technology of pilots' helmets. lite hypersonic warplane pilots. Cocooned in gel in constrictor suits to massage their blood circulation and minimise blacking out while maneuvering, pilots couldn't move a finger to control their planes. They flew by thought and imagery.

Smart protoplasmic cords of mega data capacity were twinned with the pilots' optic cords—just as with myself now, and with Miranda. Fitting our own cords, thin white threadlike worms, had involved only the most minor intrusion into my eye sockets, and hers. Once inserted, the cords found their own way, establishing their own retinal and neuronal connections.

None of this would have happened if a certain hyperjet hadn't crash-landed at its base in Nevada six years ago. If the pilot's father, a Colonel Patterson, hadn't been base commander. If his critically injured son hadn't been jammed in the crumpled cockpit in a particular position. If the Colonel himself hadn't been fitted for fly-by-thought. If for love of his son he hadn't risked an imminent inferno to say farewell. If he hadn't stared into the optics of his dying son's helmet...and suddenly received his son's mind into his own head.

Or his son's soul, as the Colonel phrased it.

Islamic countries banned guesting. Buddhists embraced it. In Britain there were tens of thousands of people in my position.

I stared into the lenses. From her side, Miranda stared. I pressed the power button. Mandalas flooded my vision, a receding tunnel of intricate light, which quickly shrank to a vanishing point as Miranda uttered a soft gasp, a sigh, nothing arduous.

Power off. Put the binoculars away in their padded case.

"Hi, Grandad," she greeted him.

If there ever *were* a way to transfer a mind into a machine I suppose we would have living tombstones to visit. Or to fail to visit. And people would have a sort of immortality.

"Maybe he'd like to see the news." Reluctant to share the start of his day with Dad, Paul switched on the TV.

The war in the Philippines, the abandonment of flooded Polynesia, a cyclone in New Zealand, minus-ninety in Alaska, a riot in the Dundee Rough, torching homes on a bitter night to keep warm…

I was thinking about my assignation (for want of a better word), and no doubt Miranda was thinking about Jenny O'Brien and an act of charity.

Maybe our daughter was a saint, an angel.

The tunnel-visions described by people pulled back from the brink of death were quite like what I and Miranda saw in the binoculars during transfer, receding in my case, approaching in hers. A couple of years ago Colonel Patterson had committed suicide. Righteous evangelists had exploited him. He'd come to believe that he'd stopped his son from going to heaven and must rectify this.

"Don't worry about the news, Grandad," exclaimed Miranda. "Everything'll be all right. Life carries on!"

We sell ourselves, is the truth of it. Lucky ones sell high, unlucky ones sell low. It seemed only logical that I should sell myself this afternoon—gratuitously and gladly—as a stage in this process. And to satisfy a burning willful curiosity.

After I'd called a couple of businesses, I messaged Denise Stuart at TV-NET. I loathed the woman in her lux job-niche, bothering anybody she cared to in the country from her superior position.

Denise came back to me presently, wearing a black kimono a-twinkle with light-emitting diodes. Long pendant crystal earrings resembled icicles.

"Ah, the spokesfemme of Omega! So we didn't have the *dernier mot* after all?"

"Denise," I said calmly, "hypothetical question. What would you say about a woman wanting to guest her dying baby? Pre-speech, pre-crawling."

Suddenly so alert. So acquisitive. "Is this true? I'd say a thousand as a finder's fee. Exclusive."

"Sorry to disappoint you!" I replied. I would sell myself in quite a different fashion, though not for nearly as much. I blanked off, and instructed my screen to reject any future calls from TV-NET.

I felt a surge of satisfaction at snubbing Denise.

The bus from our wire-fenced ville crossed a large stretch of Rough on its way to the ville where the Meridian was.

Through the window grilles of the bus I gazed at the shanties and tents, almost as medieval as those Breughel scenes back home in the toilet. Yet oddly picturesque, too. The bright patchwork clothes. The ragged children. The mongrel dogs being led about on strings—a stringless mutt would soon end up barbecued. Shebeens and bonfires. Derelict cars and vans, converted into homes. Pickers at a refuse tip hopping about like crows. A steel band playing by the roadside as if coins might spill from passing vehicles.

Decorative! A frilly, filthy collar around the neck of most villes. Suggestive of some sort of self-expressive freedom. Freedom from finance. Freedom to shiver and become slim (or grossly fat) on government-issue diet packs.

A truckload of soldiers followed our bus part of the way. The men had those multi-guns which can fire either explosive shells or humane rubber bullets or gag-gas, laugh-yourself-sick.

Then we were in semi-open country. Electric-fenced sheep pastures and pig wallows, muddy as a Somme. Fenced forestry plantations. A huge shallow lake full of trout, a watchtower upon a tiny island in the middle. By night infrared motion detectors would switch on floodlights if anything larger than a fox approached the water's edge.

Soon we entered another Rough. A solitary girl with stringy hair and a tattered false-fur coat and mitts and one of those Russian-look hats hailed us, holding up her fare and ID for the driver to see. A satchel over her shoulder.

As she made her way to the back of the bus, she tore open her coat so that she wouldn't overheat.

She smelled of patchouli, to mask body odour which the warmth of the bus quickly liberated. She grinned at me. Opened her satchel.

Bracelets and necklaces of intricately hammered tin, really quite exquisite work, I thought. If Dad had been here with me, would he have praised her? He might have enjoyed this excursion.

"Only twenty each," she said to me, meaning ten.

"Honestly I can't afford any," I told her. "Insurance, mortgage, you know."

She didn't know; though at the same time maybe she did. I was a ghost to her, of once-upon-a-time, of a maybe-world her parents may once have inhabited.

"Don't worry," she consoled me. "Women worry; men spend."

This seemed untrue, yet at least it served as a handy excuse. I hoped she wouldn't disembark at the Meridian to wander its car park hawking her craftwork to foreign businessmen. Ethnic English art, mein Herr? Monsieur? Danasama?

"Don't fret yourself, Smooth Lady."

I'd dressed confidently for my upcoming encounter. Under my thick scarf and padded coat was a high-collared side-slit shimmer-dress, revealing glittery spider-web leggings. Black pixie boots on my feet. Elbow-length black lace gloves. All of which had been packed away for years in a drawer. I'd also glamoured myself with a couple of sultry bruise-look blushers from years gone by, which hadn't dried up, having been sealed in a bag.

We soon reached the ville-fence and checkpoint. Suburbs glistened with frost which still hadn't melted, the houses like neat displays of cakes in some enormous shop. The cars cruising about seemed such shiny toys after the derelicts in the Rough.

Blind people could never become guests. Thanks be that Dad never lost his sight—that molten metal never splashed into his eye while he could still wield a welding torch!

The Rough girl stayed on the bus when I rose to leave. She would be aiming for the shopping mall—or at least for the outside of it, where the security personnel wouldn't bother her. Maybe they would even let her inside, her ID must be so clean.

When I stepped out of the bus, from the airport beyond the trade centre a silver dart was lifting into the sky, going to some place I would never go to. Dragged along at Mach 4 by its own shockwave, the hyperliner could reach the lands of orchids almost before Paul would be back from work. Its foreign destination might be abominable, crowded with beggars and refugees. Even so, a powerful airborne serenity seemed to stay with me.

The bronzed glass of much of the Meridian Hotel had discoloured so that oily pools seemed to float vertically.

In the spacious mock-marble lobby a group of Chinese or Koreans in creaseproof smart-suits were conferring with some British counterparts who looked crumpled and cheap, even though they were the dudes of the local Smooth. This was the fault of the Brits' faces, so blotchy and irregular compared with the smooth creamy features of the Asians. My compatriots' hair, even styled, was so haystacky next to wiry, trim black oriental hair.

I lingered by the menu board outside the restaurant, pretending interest, waiting for my watch display to edge closer to two. At the top of the menu a salmon leapt up a frothing spillway of fractal water to escape from the claws of a lobster, fell back, leapt up again. Liquid-crystal prices flickered as if unable to believe themselves.

When I took off my scarf and padded coat, a couple of the Asians gazed at my vamp-Vietnamese outfit appreciatively, even anticipatively, as though I might be the clincher of a deal.

Keeping my coat with me, out of a sneaking fear that it might be stolen if I left it on the rack by the porter's desk, I headed for the elevator to ascend.

Room 323. I buzzed. The lock clicked open. It would be programmed to admit one visitor, then secure itself. I pushed, and stepped into a dim bedroom. Closed curtains leaked wintry daylight. Most of the light came from the illuminated bathroom.

A short tubby man with curly black hair rose from the single armchair, barefoot, dressed only in shirt and trousers. Even in the mellow light he didn't look much like the German I'd imagined.

"Herr Schmidt?"

"That is I, dear lady." Nor did his accent have the perfection of most Germans speaking English.

I must have looked bewildered, for he proceeded to explain himself, quite proudly. His parents had been Turkish guest-workers, but he was German and had changed his name accordingly. He had prospered; he was his parents' success story, a full European. Manfred Schmidt, who might once have been Mustafa.

When not in Germany he could allow himself to be somewhat Turkish in a playful fashion. This took the form of jokes about harems and slave-girls—of whom I was now an honorary embodiment for an hour—and silly proverbs.

"A woman possesses a precious candlestick," I learned, "but the man has the candle!" And Manfred-Mustafa's candle needed attention.

In fact he was quite sweet and gentle—"A lion does not harm a lady"—though I wouldn't have wished to be married to him.

"Open up for me like a marrow flower," he commanded.

I imagined myself as an orchid instead—a soft, lush, velvety orchid being assaulted for nectar by a magically hovering hyperliner, wings fluttering as fast as his heartbeat, and soon my own heartbeat too.

I even experienced shockwaves, which surprised him.

"A hen cannot live without a cock," was his opinion. By then his candle was quenched.

"A man has one desire," he confided sadly, "yet a woman has nine."

✦

Afterwards, dressed again in my vamp-Viet gear, I left him, carrying my coat over my arm on account of the warmth, my black lace gloves stuffed in the pocket.

I knew at once it was Paul who came out of the room closest to the lift. He was whistling tunelessly in a show of nonchalance, in case any maid was in the corridor.

Rage. I felt such rage.

"Is she still in there?" I demanded.

He jerked. He didn't recognize me instantly in my slit dress and pixie boots and blushers.

"No—" Only then did he realize. "Cath—!" How he blanched.

Already my anger had drained away. I could almost see the thoughts tumbling through his mind. He was like some gambling machine gone faulty. It just couldn't halt its reels on any pay-out line. He was thinking that I'd suspected him of infidelity and somehow I'd trailed him to catch him out. Worse: I suspected him of unfaithfulness subsidized by me sitting all day in front of my Omega screen.

Would it be best if he invented some fictitious girlfriend and swore never to see her again? No, that wouldn't wash—I would demand details and more details until his fib fell apart in tatters.

Yet why was I all vamped up? Was this my disguise?

"How did you—?" he asked.

"How often—?" I asked at the same moment. We seemed to speak in the shorthand of those who have known each other for a long time.

Manfred-Mustafa would still be showering, sprucing himself. He wouldn't be stepping outside for five or ten minutes to attend to whatever business brought him here. I had the edge on the situation.

"How often, Paul?"

"Not often," he protested. He hadn't the nerve to claim that this was the first and only time. "Sam Henson sometimes gives me a bit of cash on the side, not in the books—"

So the gambling machine had selected a line-up featuring… money, of course. Me slaving in front of a screen. Him squandering. Cheating on the family budget. We *would* think that way nowadays, wouldn't we?

I played along. By now it was quite impossible for him to accuse me, tit for tat, unless we wanted to ruin ourselves—and ruin Miranda's prospects—by separating.

"You're diddling the discount—"

"Or something," he agreed.

Creative accountancy is the lifeblood of the motor trade. Selling cars to one's own company to meet quotas. Marking trade-ins at imaginary values. Insurance and finance kick-backs.

He gawked at me, unsure that he had chosen the appropriate pay-off line but unable to spin the reels again.

"Never again," he promised. And I nodded. Actually, I could almost have laughed. Yet my laughter might have become hysterical.

"Well," said I, "what a surprise."

It wasn't a total surprise—apart from the coincidence of our meeting here, and that wasn't really too amazing. Was Paul a customer of Mrs Appleby or was his go-between somebody else?

"I'm going home now," I said. "You'd better get back to work."

"I'll never do this again," he vowed. "It's just that—"

"No, it isn't *just*."

For him to chauffeur me home through the Rough—presuming that he'd driven to the Meridian in a Jag—would have been as impractical as it would have been absurd. The display model would need to be back toot-sweet in the armoured-glass showroom.

✦

When Miranda came home just ten minutes after me, she was more tired than usual, but effervescent.

"We talked to Jenny O'Brien—didn't we, Grandad? It seemed to help. How were the orchids, Mum?"

"Indescribable," I said, quite truthfully. "All the heat they need to be so beautiful!"

We transferred Dad back into me. A day at school seemed to have done him a power of good. He settled back contentedly, deep inside me.

When Paul arrived home, he was extremely unsure of his reception.

"Denise Stuart called me this morning," I said to him, "but I told her to fuck herself."

Miranda gaped.

"You need to be firm with some people," I told our daughter. "Without upsetting applecarts unnecessarily! Without rocking the boat so that it sinks."

"That's true," Paul agreed feebly.

And I said to him, in a mock-foreign accent, "When a husband comes home, a wife should count his teeth." Actually, what Manfred-Mustafa had quoted was exactly the opposite. No man should allow his wife to count his teeth; to know how much money he has.

Paul stared at me weirdly. Did he imagine I was proposing he should make love to me tonight?

"The marrow flowers open wide," I said.

This too might seem suggestive. Yet I was thinking a few weeks ahead—to when I might reasonably ask Miranda to look after Dad for another day.

I would insist on a foreigner. Any foreigner. Not one of my own countrymen. The money tyranny had been getting worse for years—

not in an ebullient American way (well, that's a ridiculous generalization!)—but selfishly and divisively, breeding a society of fear and of foolish hopes, until it finally destroyed the soul of the country.

Being able to harbour a loved one's soul or mind, all those acts of charity and sacrifice, might have seemed an exception. It damned well wasn't.

Only foreigners, Mrs Appleby, only foreigners…! Of course they would have their own problems. Even including guests. Might I ever meet a man who would invite a guest to join in enjoying me?

The marrow flower opens wide.

Paul was bewildered but he knew better than to ask what I meant.

ATTACK OF THE CHARLIE CHAPLINS

BY GARRY KILWORTH

SCENE I

A subterranean bunker somewhere in South Dakota. Feverish activity is taking place within the confines of the bunker. In the centre of it all a middle-aged general is musing on the situation which unfolds before him.

Reports are coming out of Nebraska that the state is under attack from heavily armed men dressed as Charlie Chaplin. My first thought was that a right-wing group of anti-federal rebels was involved. It seemed they were using irony to make some kind of point. After all, Charlie was eventually ostracised to Switzerland for having communist sympathies.

As more accurate reports come in however, it becomes apparent that these are not just men dressed as Charlie Chaplin, they are the real McCoy—they are he, so to speak.

"It's clear," says Colonel Cartwright, of Covert Readiness Action Policy, and the Army's best scriptwriter, 'that these are aliens. What we have here, general, is your actual alien invasion of Earth. Naturally they chose to conquer the United States first, because we're the most powerful nation on the planet."

"Why Nebraska, colonel?" I ask. I am General Oliver J.J. Klipperman, by the way. You may have seen my right profile next to John Wayne's in The Green Berets. I was told to look authoritative, point a finger in the direction of Da Nang, but on no account to turn round and face the camera. I have this tic in my left eye and apparently it distresses young audiences. 'Nebraska isn't exactly the most powerful state in the Union. Why not New York or Washington?"

Cartwright smiles at me grimly. 'Look at your map, General. Nebraska is slap bang in the middle of this great country of ours. It has one of the smallest populations. You get more people on Fifth Avenue on Christmas Eve than live in Nebraska. You simply have to wipe out a small population and you control this country's central state. Expand from there, outwards in all directions, and you have America. Once you have America, you have the world. It's as easy as that."

I nod. It all makes sense. Nebraska is the key to the control of the U.S. of A. The aliens had seen that straight away.

"What do we know about these creatures?" I ask next. 'The President will expect me to sort out this unholy mess and I want to know who I'm killing when I go in with my boys."

The colonel gives me another tight smile. 'These creatures? Nothing. Zilch. But we have a trump card. We've been preparing for such an invasion for many, many years and our information is voluminous."

"It is?" I say. 'How come?"

"Hollywood and the Army connection," says the colonel. 'Army money, personnel and expertise have been behind every alien invasion movie ever made."

"It has?" I reply. 'I mean, I knew we had fingers in Hollywood pies—I've been an extra in over a dozen war movies—but every alien invasion movie? Why?"

[Zoom in on colonel's rugged features.]

'Training," the colonel says, emphatically. 'Preparation. If you cover every contingency, you don't get surprised. We've been making films of alien invasions since the movie camera was first invented. We've covered every eventuality, every type of attack, from your sneaky fifth column stuff such as Invasion of the Body Snatchers to outright blatant frontal war, such as Independence Day. We know what to do, general, because we've done it so many times before, on the silver screen. We know every move the shifty shape-changing bastards can make, because we've watched them in so many films. Alien, War of the Worlds, The Day the Earth Stood Still, Close Encounters of the Third Kind, you name it, we've covered it. On film." He pauses for a moment, before saying to himself, 'That speech is a little long—I'll have to think of some way of cutting it when we make the actual movie of this particular invasion."

[Back to middle-distance shot.]

Something is bothering me. I put it into words.

"Weren't they friendly aliens in Close Encounters?"

"No such thing, general. What about those poor guys, those pilots they beamed up from the Bermuda Triangle in December 1945? They kept them in limbo until their families were all dead and gone, then let 'em come back. Is that a friendly thing to do?"

"I guess not. So, colonel, we've had all these exercises, albeit on celluloid, but what have we learned? What do you suggest we do with them?"

"Blast them to hell, general, begging your pardon. If there's one thing we've learned it's that if you give 'em an inch, they'll take a planet. They've got Nebraska. That's almost an inch. We need to smash them before they go any further. Blow them to smithereens before they take Kansas, Iowa or Wyoming or, God forbid, South Dakota."

I always err on the side of caution, that's why I'm still a one-star general, I guess.

"But what do we actually know about these creatures? I mean, why come down here looking like Charlie Chaplin?"

The colonel's eyes brighten and he looks eager.

"Ah," he says, 'I have a theory about that, sir. You see, we send crap out into space all the time. I don't mean your hardware, I mean broadcasts. They must have picked up some of our television signals. What if their reception had been so poor that the only thing they picked up was an old Charlie Chaplin movie? What if it was one of those movies in which he appears on his own—just a clip—and, here's the crunch, they thought we all looked like that?"

The colonel steps back for effect and nods.

"You mean," I say, 'they think the Charlie Chaplin character is representative of the whole human race?"

"Exactly, sir. You've got it. We all look alike to them. They came down intending to infiltrate our country unnoticed, but of course even most Nebraskans know Charlie Chaplin is dead, and that there was only one of him. The dirt farmers see a thousand look-alikes and straight away they go, "Uh-huh, somethin's wrong here, Zach..."

"So they did what any self-respecting mid-Western American would do—they went indoors and got their guns and started shooting those funny-walking little guys carrying canes and wearing bowler hats."

"I see what you mean, colonel. They're "not from around here" so they must be bad guys?"

"Right."

"Blow holes in them and ask questions later?"

"If you can understand that alien gibberish, which nobody can."

"I meant, ask questions of yourself — questions on whether you've done the right and moral thing."

"Gotcha, general."

I ponder on the colonel's words. Colonel Cartwright is an intelligent man—or at least what passes for intelligence in the Army—which is why he is a senior officer in CRAP. He hasobviously thought this thing through very thoroughly and I have to accept his conclusions. I ask him if he is sure we are doing the right thing by counter-attacking the aliens and blowing them to oblivion. Have they really exterminated the whole population of Nebraska?

"Every last's mother's son," answers the colonel, sadly, 'there's not a chicken farm left."

[Gratuitous shot of a dead child lying in a ditch.]

'And we can't get through to the President for orders?"

"All lines are down, radio communications are jammed."

"The Air Force?" I ask, hopefully.

"Shot down crossing the State line. There's smoking wrecks lying all over Nebraska. Same with missiles. We were willing to wipe out Nebraska, geographically speaking, but these creatures have superior weapons. We're the nearest unit, general. It's up to us to stop them."

"How many men have we got, colonel?"

"A brigade—you're only a brigadier-general, general."

"I know. Still, we ought to stand a chance with four to five thousand men. They...they destroyed our whole Air Force, you say?"

[Zoom in on TV screen showing smoking wrecks.]

The colonel sneers. 'The Air Force are a bunch of Marys, sir. You can't trust a force that's less than a century old. The Army and the Navy, now they've been around for several thousand years."

ATTACK OF THE CHARLIE CHAPLINS

There had never been much call for the Navy in Nebraska.

[Back to half-frame shot.]

'Are we up to strength?"

"No, sir, with sickness and furlough we're down to two battalions."

"Okay' I state emphatically. 'We go in with two thousand, armour, field guns and God on our side."

"You betcha!"

[Enter Army corporal carrying sheet of paper.]

'Yes, corporal?" I say, icily, recognising her as the extra who upstaged me in the remake of The Sands of Iwo Jima by obscuring my right profile with her big knockers. 'I'm busy."

"I thought you ought to see this message, sir." She offers it to me. 'Just came through."

"From Washington?" I ask, hopefully.

"No, sir, from the alien."

"The alien?" I repeat, snatching the signal. 'Youafraid of plurals, soldier?"

"No, sir, if you'll read the message, sir, you'll see there's only one of him—or her."

The message is: YOU AND ME, OLIVER, DOWN BY THE RIVER PLATTE.

"Looks like he's been watching John Wayne movies, too," I say, handing Cartwright the piece of paper. 'Or maybe Clint Eastwood."

The colonel reads the message. 'How do we know there's only one?" he asks, sensibly. 'It could be a trick."

"Our radar confirms it, sir," the corporal replies. 'He's pretty fast though. It only looks like there's multiples of him. He seems to be everywhere at once. He's wiped out the whole population of Nebraska single-handed."

"Fuck!" I exclaim, instantly turning any movie of this incident into an adult-rated picture. 'What the hell chance do I stand against an alien that moves so fast he becomes a horde?"

"Fifty percent of Nebraska was asleep when they got it," says the colonel, 'and the other half wasn't awake."

"What's the difference?"

"Some of 'em actually do wake up a little during daylight hours."

"You think I stand a chance?"

The colonel grins. 'We'll fix you up with some dandyhardware, sir. He'll never know what hit him."

"But can I trust him to keep his word? About being just one of him? What if he comes at me in legions?"

"No sweat, general," says the colonel. "This baby—

[Close-up of a shiny gismo with weird projections.]

"—is called a shredder. Newest weapon off the bench. One squeeze of this trigger and it fires a zillion coiled razor-sharp metal threads. Strip a herd of cattle to the bone faster than a shoal of piranha. You only have to get within ten feet of the bastard and you can annihilate him even if he becomes a whole corps."

"Can I hide it under my greatcoat?"

"Nothing easier, sir. And we'll wire you with a transmitter. He's only jamming long-distance stuff. You can tell us your life story. Oh, and one more important thing."

"What's that?"

"We have to give him a nickname, general."

I stare at the colonel. 'Why?" I say at last.

"Because that's what we're good at. We always give the enemy a nickname. It demeans them. Makes them feel self-conscious and inferior. It's our way of telling them that they're the lowest form of human life."

"Or, in this case, alien life."

"Right, general. So we have to give him a humiliating nickname—like Kraut, Slopehead, Raghead, Fritz, Dink or Charlie…"

"We can't nickname him Charlie, he's already called Charlie."

"Okay, I take that on board. How about we call him Chuck?"

'Doesn't sound very demeaning to me. My brother was called Chuck."

"Depends on how you say it, general. If we're talking about your brother, we say "Chuck" in a warm kind of tone. But if we're talking about Chuck, we use a sort of fat, chickeny sound—Chuck—like that."

"I think I understand, colonel. Well, let's get me armed and wired. It's time I taught Chuck a lesson."

Stardom here I come. A part with lines. My part. A lone, courageous part, if they let me play myself in the movie—providing I live to rejoice in it, of course.

SCENE 2

Somewhere out on the plains of Nebraska. A man is walking down towards the River Platte. The night is dark but studded with bright stars, giving the impression of vast distances and emphasising the insignificance of the brave lonely figure. The brave lonely figure is apparently talking to himself.

Are you listening back there in the base? The moon is gleaming on my path as I reach the banks. Here in the humid Nebraskan night I wait for my adversary. Single combat. Mano a mano. The old way of settling differences in the American West.

Hell, what am I saying, we didn't invent it. The old, old way. The chivalric code of the knights. A tourney. A duel. An affair of honour. Rapiers at dawn. Pistols for two, coffee for one.

[*Aside: We're kind of mixing our genres here with Westerns and Science Fiction, but I think we can get away with it since the two have always had a close relationship, being drawn from the same source—the conquest of frontiers by American pioneers.*]

And I am ready. You didn't send me out unprimed, colonel. You made me submit to brainstorming. Masses of data has been blasted into my brain in the form of an electron blizzard. Every extraterrestrial invasion movie ever filmed is now lodged somewhere inside my cerebrum, waiting to be tapped. Any move this creature makes, I'll have it covered. Hollywood, under secret Army supervision, has foreseen every eventuality, every type of Otherworlder intent on invading and subduing us Earthlings. They're all in my head.

[*A solitary charred and wounded chicken crawls silently across the landscape.*]

Swines! Uh-huh. More movement up there.

Chuck's coming up over the ridge! Thousands of him doing that silly walk with the cane and twitching his ratty moustache. This is really weird. A swarm of Charlie Chaplins. Did he lie? Is he going to come at me in hordes? Boy, can he move fast. They're all doing different things. One's swinging his cane and grinning, another flexing his bow legs, yet another pretending to be a ballet dancer. Multitudes of him, pouring over the ridge now, like rats being driven by beaters.

"Don't let him get to you with the pathetic routine," you warned me, colonel. 'You know how Chuck can melt the strongest heart with that schmaltzy hangdog expression. Don't look at him when he puts his hands in his pockets, purses his lips, and wriggles from side to side." Well, don't worry, I hate Charlie Chaplin. That pathos act makes me want to puke, always did. If he tries that stuff, I'll shred him before he can blink.

ATTACK OF THE CHARLIE CHAPLINS

He's getting closer now, moving very slowly. He's suddenly become only one, a single Charlie Chaplin. I can see the white of his teeth as he curls his top lip back.

My fingers are closing around the butt of the shredder. I'm ready to draw in an instant. The bastard won't stand a chance. Wait, he's changing shape again. Now he's Buster Keaton. I never liked Buster Keaton. And yet again. Fatty Arbuckle this time. I detest Fatty Arbuckle. Someone I don't recognise. Now Abbot and Costello. Both of them. The Marx Brothers.

Shit, he's only eleven feet away, and he's changing again. He's gone all fuzzy. He's solidifying. Oh. Oh, no. Oh my golly gosh. God almighty. It's…it's dear old Stan Laurel. He's got one hand behind his back. I guess he'sholding a deadly weapon in that hand.

"Hello, Olly."

Did you hear that, colonel? Just like the original. He…he's beaming at me now, the way Laurel always beams at Hardy. And I…I can't do it. I can't shoot. He's scratching his head in that funny way of his. Of all the comic actors to choose. I loved Stan Laurel. I mean, how can you shoot Stan Laurel when he's beaming at you? It's like crushing a kitten beneath the heel of your boot. I can't do it. The flesh may be steel, but the spirit's runny butter.

Tell you what I'm going to do—I'll threaten him with the shredder. That ought to be enough for Stan Laurel.

Oh my gosh, he's burst into tears.

"Don't point that thing at me, Olly. I don't want to hurt you. I just want to be your chum."

I've put the weapon away. He's smiling again. He's offering me a cigar. Hey, you should see this, colonel. He's done that trick, you know, flicking his thumb out of his fist like a lighter? There's a flame coming from his thumbnail.

He's still smiling. He's friendly after all, though he's still got one hand hidden from me. Maybe he's realised he's made a mistake? I have to show willing. I'm taking a light for the cigar. Hell, he could be a really nice guy.

I don't know what this thing is, but it's not a Havana cigar. Tastes kinda ropey, like the cigar that producer of a low-budget B movie once gave me, when I played Young Ike. What was his name? Ricky Hernandez, yeah. Good movie that. Pity it was never released.

Jesus, this thing is playing havoc with my throat. Can you still hear me? It has a familiar smell—now where did I—oh yes, in the Gulf. Shit, it's nerve gas! The bastard has given me a cigar which releases nerve gas into the lungs.

Oh fuck, oh fuck—I'm getting dizzy. I feel like vomiting. There's blood coming from my mouth, ears and nostrils. He's reaching forward. He's taken my weapon. I'm… I'm falling…falling. Oh God, my legs are twitching, my arms, my torso, my head. I'm going into a fit spasm. I'm dying, colonel. I'm a dead man.

Wait, he's standing over me. I think he's going to speak. Are you listening, colonel?

"You're supposed to say, "Another fine mess you've gotten me into, Stanley," and play with your tie."

Hollywood, damn them. He's speaking again. Listen.

"I suppose you think I lied to you, Olly?"

Yes I do, you freak, you murdering shape-changing bastard. I do think you lied to me.

He's giving me one of those smug Stan Laurel smiles, showing me his other hand, the one he's had behind his back all the time. He's…he's got his fingers crossed.

"Sorry, Olly."

Hollywood covered every contingency except one. In all the alien invasion movies they ever made, the attacking monsters are always as grim as Michigan in January. As I lie here dying, the joke is on you and me, colonel. There's one type of extraterrestrial we didn't plan on. An offworlder just like our own soldiers.

An alien with a sense of humour.

FADE OUT

FOR LIFE

BY CHRISTINE MANBY

"So are you going to buy anything for me when you go shopping, Sugar Mummy?" came a babyish voice from the shadows of the bedroom.

"What do you think you deserve?"

"Oh, I don't know. Surprise me. I'll be pleased with anything."

Persephone Rayfield smiled. Money could buy you just about everything except, as ever, love. This was something she knew only too well as she sat on the balcony of her Malibu home and pondered the fate of a forty-million-dollar bonus from another great year on Wall Street.

Before her on the table was a neatly written list of options to blow the lot. The latest New Generation Ferrari? No. They'd lost all their grunt since the emission laws restricted so much as a public fart. An entire wardrobe of antique Armani? She would only ever be able to wear two hundred thousand dollars' worth at a time. Perhaps a fabulous diamond necklace that would cover her entire breastplate? And get her head severed from her neck as she made the short journey from the night-club to the car.

Persephone put a single straight line through each of the glittering options and tried to start again. Had shopping always been this difficult? she asked herself. She looked down at the surfers riding the waves near the beach below. Simple pleasures. Though only experts and idiots would tackle surfing since French nuclear testing on the moon had messed up all the tides.

"How about a new car?" asked her companion. "A Ferrari?"

"Already persuaded myself against that."

"A new home cinema system? You could get a 3D screen."

"There's more to watch from my window."

Persephone picked up her antique binoculars to get a closer look at one surfer who was doing particularly well. He reminded her of the Roman statues that had been washed into the sea from the Getty Museum when it finally fell from its cliff. As she watched, the surfer ducked into a tube and disappeared for what seemed like an eternity until the ocean spat him out again at the other end in a splutter of sea foam.

"You're so difficult to please. You don't know what you want."

"Oh no," replied Persephone. "I'll know exactly what I want when I see it."

For example, she wouldn't have said "no" to the surfing boy.

Persephone already knew that his name was Peter and he belonged to her friend Serena Strane. She could see Serena walking up to him now, holding out a fluffy towel to wrap him in when he finally gave up on the waves. Peter was gorgeous. Tall, blond, bronzed in that old-fashioned way. (Though not from the sun of course. No one ever went out in the Californian sunshine without a total zinc body block anymore. It was all done with extract of carrot now, as it had been in the 1970s. Funny to think that all the self-tanning technology of the late twentieth century had been proven to be more harmful than the sun itself.)

Peter was rumoured to be quite bright too. Educated in England just before the Gaian fundamentalists there decided to impose the "no education for men" law they had found in the latest good book. Serena had been incredibly lucky, getting Peter into California before the immigration laws were tightened up and the gates to the east started to close. She had got herself all that brawn and a brain and still enough change from two million bucks to retile her entire home. Her entire sprawling home. You needed a phone to talk to someone at the other end of the breakfast bar in her Santa Monica kitchen.

"What are you looking at out there?" Her lover was frustrated by the lack of attention.

The glint of the sun on the glass of Persephone's binoculars drew Serena's attention up to the hills. There was no way that Serena would be able to see Persephone at such a distance, unless she was wearing those incredible zoom contact lenses of hers again, but Persephone knew that Serena had guessed who was watching. She waved. But there was no malice in her wave. Serena had something that very few people could afford, and as such she felt that she was not only Peter's keeper but his curator. It was her duty to share the joy of such rarity with the other girls. Albeit on a look-but-don't-touch basis.

Persephone put down her binoculars with a smile and added another word to her shopping list.

"Have you made your mind up yet?" came the whiny voice from the bedroom again. "I want to go shopping with you before everything's closed."

"I've nearly finished, darling." Persephone wrote out a cheque and rolled it into a little tube. It was a small amount, but she had written it in big letters. "Come here, you nuisance," she said now, beckoning her young lover towards her, out in the weak afternoon sunshine. Pip appeared instantly and curled into an appealing ball at Persephone's feet to wait for a sign of her affection.

"What have you been writing?" Pip asked. "Can I see?"

"It's private."

"Is it what you're going to buy with your bonus?"

"Could be, sweetheart. If I can still get hold of one."

"Get hold of one? What is it?"

"It's a surprise."

"Can I use it too? Or are you getting me a separate present?" Pip pouted in the way Persephone might have found endearing just half an hour before.

"You can choose your own present," said Persephone. "Take this bit of money here and buy something for yourself." She tucked the rolled-up note into the cleavage of Pip's silky dressing-gown. Pip kissed her excitedly, not bothering to check the amount.

"You're so generous."

"Sssh. Hurry along," Persephone said with a hint of embarrassment at Pip's pathetic gratitude.

"I love you, Persephone. I'll see you later."

Persephone would have changed the code on the entry-gate before the girl got back.

She hoped that Pip would realise it wasn't because of anything she had done. Pip was sweet enough. A beautiful girl. But then all the girls in Hollywood were beautiful. Popped out of the same mould. Though their bodies were perfect, they held only as much interest for Persephone as the dolls they resembled. Now Persephone had enough money to try something different. To buy something different.

She looked once more at the new word on her shopping list.

Man.

Why not?

"You cannot be serious," Angela, her best friend, said later that night. "It's immoral."

"Why?" asked Persephone defensively.

"Because you can't buy another human being. Even if it is only a man."

"I won't be buying him. I'll be saving him from a terrible life. Can you believe that in some parts of China they actually leave baby boys to die in those dreadful cold, dark rooms? They just let them starve to death to avoid the economic burden of bringing them up."

Angela raised an eyebrow. "Oh, so it's a mercy mission. You're getting a Chinese boy?"

"Well." Persephone blushed. "Not exactly."

"Ha."

"The red tape in China is absolutely unbelievable," Persephone said by way of an excuse. "Even if you get to the stage where you're handing over the cheque, there's still no guarantee that you'll get the goods."

"The goods," Angela repeated scornfully.

"You know what I mean."

"You could just donate all the money you're going to spend to that charity that's setting up homes for those poor children overseas."

Persephone spluttered into her glass. "No way. I did that auction for the homeless hostel last year."

"And now you've got compassion fatigue."

"Give me a break. I just want to have a bit of fun. I could have got a Ferrari instead, you know. I wouldn't have bothered asking you for advice then."

"And what fun would you have had with an average speed of just three miles an hour in the city? I'm glad you saw ecological sense."

"Yeah, you see," said Persephone, seizing the opportunity to get back into Angela's good books. "I've already thought about my duty to the Earth. Now I need to think about my duty to me. Besides, don't tell me you've never considered it. Not even when you got that huge inheritance last year?"

"I did think about, I admit. But, thank goodness, I realised the stupidity of it all within three seconds and bought that Real-Time Intimacy Machine instead."

"Ah yes. The RIMMER." Persephone smirked as she tried to suppress an image of Angela touching herself up in twenty million dollars' worth of superfine carbon-fibre suit.

"The brochure refers to it as the 'luxury RTI' actually," Angela replied.

"If you say so. But things are different for you, Angie. You're in a relationship. You get everything you need from Jennifer except for the obvious; and now you've got the 'luxury RTI,' she can pretty much give you that as well."

Angela sighed. "There is such a thing as a clitoral orgasm, you know. You've just spoiled yourself with your maniacal use of that Superskin dildo I bought you for your birthday."

"If you'd ever tried it, you would understand. It feels real."

Angela's face twisted into a sly smile.

"Oh God," Persephone winced. "You did try it, didn't you? Before you gave it to me?"

"No, I did not," Angela protested. "But anyway, how do you know if it feels real or not?"

"Exactly," said Persephone.

"Exactly," Angela repeated with a very different emphasis on the word. "What you think feels 'real' might actually be far and away better than the real thing. How about that?"

"I'm not convinced."

"Persephone." Angela put on her serious face. "What do you want to get a man for?"

"Because I want something to stroke, and I can't get a dog."

Angela snorted.

"I'm going to get an English one," Persephone elaborated. "One with some class, pedigree and breeding. They're easy to get hold of. Since the Gaian fundamentalists took over the government there, the men have been banned from work and they're just dying to get out."

"Yes, dying by the laser as they try to get over the border into Wales. Sure, you'll find hundreds that want to come over here but how on earth will you get him a visa?"

"How do you think I'll get him a visa? I'm paying forty million bucks. You know, Angela, I just can't understand it. Here we are in California with hardly any men at all since the plague, and there are all these fundamentalist states springing up around the world, rendering their menfolk useless with dogma but refusing to let us take them off their hands."

"Men were useless to us long before the rise of the Gaians," Angela said sagely. "I can't believe you want to lumber yourself. Sure, we all lost a few loved ones to the plague, but in return we got our freedom. I mean, most of the women in our grandmothers' generation, even though they had Margaret Thatcher and Madonna as role models, were as shackled to the kitchen sink as women two thousand years before them. And they would never have dared dabble in a same-sex relationship. Now you can be whatever you want to be. And you can be with whomever you want while you're being it."

"As long as they're a woman," Persephone sniffed.

"Well, yes... But I can't believe you're seriously hoping for a relationship with a man," Angela continued in the face of Persephone's dissent. "For a start, any man you meet through an agency will simply be after a ticket out of the UK. And more importantly, men just can't love like women can. It's no wonder they got the dog virus, because they are like dogs. They just follow their basest instincts. They go where the food and the sex are."

"You're being unfair."

"And you're being stupid. If you spend your bonus on a man, it won't stop there. There'll be medical bills and insurance for a start. And you'll still have to buy that bloody Ferrari so that he can rev it like he's planning to go into orbit every time the traffic moves forward half an inch."

Persephone leaned over the side of her chair and pulled an untitled folder containing sheets of badly copied photographs out of the magazine rack. "If you take a look at these, I'm sure you'll change your mind."

Angela flicked through the pictures.

"Aren't they beautiful?"

"In a traditional kind of way," Angela conceded. "They're all straight lines, aren't they? Hard corners. It's like 'look at me firing on all cylinders' just because they've got dicks. It's old fashioned, sister. Get yourself a new girlfriend." Angela folded the brochure shut and handed it quickly back to Persephone, though her eyes lingered on the cover picture just long enough for Persephone to notice that there might have been a spark of interest there.

"If I get one, you can have a ride. Since you're my very best friend."

Angela laughed as she stood up to go. "Don't be disgusting, Persephone. I have a wife at home. Call me when you want to know more about the RTI."

But Persephone was not about to be put off by arguments about the expense and ethics of it all. Lesbianism may have been the state-approved relationship style but Persephone could never shake off the feeling that she was missing out on something.

She had felt like this for as long as she could remember and while she had the money in her pocket she couldn't hope to put the thought out of her mind. Besides, she told herself, she would be doing someone a favour. California was becoming more and more isolated from the rest of the world day by day, particularly since the big quake that made the roads to the rest of America pretty much impassable. If she waited even a week, Persephone panicked, she might lose the opportunity to scratch her life-long itch altogether. Plans to bring in instant DNA testing at immigration, to prove that incoming foreigners were really related to the people they said they were visiting, had already been approved by the Senate and there was talk that the tests might even be applied to Americans from other states. With that in mind, Persephone went to the agency that had done so well for Serena the very next day.

Persephone had never been so far downtown before. With the prices they were charging, she would have thought that "Intermates International" could afford to have offices on Rodeo Drive. In fact, they had once had offices at a much smarter address, but since the new immigration laws, these places were effectively illegal. And so, although their charges had soared by hundreds of percent in the two years since Serena had purchased her Peter, their premises had grown progressively seedier with each price hike. They needed to be inconspicuous, able to pack up and disappear overnight. Twenty years earlier and Persephone could have done all her correspondence on the Internet under the watchful eye of a body set up to make sure she wasn't ripped off. Now, as Angela had warned her, being put in touch with a man was as dangerous as it had once been to score some crack.

The woman at "Intermates International" didn't look as though she had ever been lucky enough to sample her own merchandise. When Persephone appeared at her office with the banker's draft for forty million dollars burning a serious hole in her pocket, she was shown through to a poky "assessment room" as though she were coming in for a visit to the dental hygienist after six years of not flossing her teeth. Had Serena's endorsement of the end product not made Persephone so determined, she might have walked out there and then and looked into buying an RTI.

"Call me when you're done," the agent said.

Persephone sat down at the dusty console and picked up the battered VR helmet which sat alongside it. It was an ancient system, hopelessly out of date. When the helmet was fastened firmly in place, the selection process began with a synthetic rendition of "Land of Hope and Glory" accompanied by a graphic of two fluttering flags. The Stars and Stripes for America and the Union Jack for the UK (though it hadn't been United since the Welsh fundamentalists blew a channel down the border with England ten years before, which confirmed the age of the VR system). After that burst of nostalgic patriotism, Persephone chose her avatar for the session. The choices were several great women from Queen Boudicca to Chelsea Clinton, ex-President of the United States, none of whom exactly touched Persephone's heart. She chose Cleopatra in the end, but not without a frisson of embarrassment. It seemed such a pretentious choice. But on the other hand, she could hardly identify with Mother Theresa when it came to the matter of choosing a man, could she?

In her electronic form, Persephone progressed into the heart of the programme. She quickly found herself in a white-washed virtual hall, with doors to the left and to the right of her, and strolled straight past all doors referring to female mates until she found the opening that she was looking for. MEN, the sign proclaimed in shimmering golden letters. Persephone's hand, connected to the circuit by a dubiously worn-looking glove that she hadn't expected to work, pushed against the unnaturally shiny handle until the door gave way... and she was in a corridor again, sighing with disappointment, as though she could honestly have expected to open the door onto a room crowded with guys.

Blond. Dark. Under six feet. Over 180 pounds. The labels on the doors categorised the candidates as simply as if they were jars full of spices. Persephone regarded these choices with bewilderment. She had planned to specify someone along the lines of Serena's Peter, but now she was scared that by making a decision about one characteristic she might be cutting out Mr. Wonderful just because he had the wrong colour hair.

Persephone's hand hesitated on the handle of the door marked blond. Historically, Latin-looking lovers were supposed to be more passionate, weren't they? As door gave way to door, Persephone realised that there must be simply hundreds of combinations of physical appearance. You could specify almost everything, right down to the number of moles on his back, though it was pretty hard to get worked up about the shape of a guy's fingernails. Persephone answered each choice with the description that she thought closest to the way Peter looked. After all, looks were secondary really.

The personality part came last.

Her interest renewed, Persephone first let herself into the corridor labelled "placid." A good quality, she thought, the kind of thing you looked for in a Labrador. From there to "humorous." She didn't want him to be too deadpan around her friends. Persephone was about to look for "romantic" when she suddenly became aware of a whirring noise in the background. The corridors quickly gave way to the flags again. The test had stopped.

Persephone hurriedly took off the glove and then the helmet. It couldn't be over, surely? She had only specified two personality characteristics. The machine must have broken down, she thought. But it hadn't. When Persephone turned around on the creaking swivel chair, she saw that the agent was standing behind her again, leaning over an ancient laser printer as if by staring at it viciously she could make it spew something out more quickly.

"Is that it?" Persephone asked the woman's back. "I mean, I didn't even get to see any pictures of the men on your books. And there weren't many questions about temperament and personality, were there?"

"Most people don't even notice that," said the agent. Persephone thought the woman's hard face almost broke into a smile when she saw what Persephone's three hours on the machine had produced. The printer spat out a kind of photofit image of the wanted man. Six foot plus. Blond and handsome. Of course.

"We don't promise a love match," the agent added pointedly. "I know how romantic you Cleopatras are. Though it is still a big commitment you're entering into. You realise that if we send you someone you're not happy with, you will still be liable for all costs incurred while getting him through immigration, including bribes, and have to pay an allowance to him until he finds an alternative sponsor? And even if he does, he might not tell you. You could go on paying him forever and never get any action."

"Of course," said Persephone.

"If you're fine with all that, we should have someone for you in about a week."

"A week?"

"Sure. There's no problem getting English men these days. They're desperate to get out of there. Country's full of fundamentalist nuts who want to have them executed for having dicks. But," she placed her hand professionally on Persephone's arm in a gesture calculated to inspire confidence, "you can rest assured that we still screen for quality despite the influx, Ms. Rayfield."

"A week?" Persephone repeated disbelievingly. "What about immigration?"

"As I said, there'll be bribes. Forty million dollars is still quite a lot of money to some people."

The agent handed Persephone her jacket and showed her to the door rather hurriedly, as though she wanted to get out of there herself. "Oh, I almost forgot," she said as Persephone was stepping outside. "We'll need a deposit."

Persephone authorised the transfer of twenty million dollars to the account of "Intermates International," which traded under the innocuous name of "Inter-office Services," and went away feeling slightly hollow.

That night, she called Serena.

"Of course you feel hollow, my dear. I'm not saying that it isn't a great deal of money but believe me, it will be worth it. You'll be the most popular girl on the block."

"What if I don't like him? I'll still have to pay for him."

"You could hire him out."

"Serena!"

"Where's your problem with that? You would lend me your car for a fee, wouldn't you?"

"Yeah, I suppose. Oh, I don't know if I've done the right thing. It would be so much easier if I could have a dog."

"No-one can get a dog. It doesn't matter how much money you've got."

"I don't know. I was reading something in the paper today. One of the left- wing senators from upstate is lobbying for the reintroduction of certain breeds. They reckon that the culling and neutering policy during the Parvo crisis succeeded in wiping the virus out. No reason why they shouldn't reintroduce some of the dogs now it's gone."

"What? It'll never happen, Perse."

"I think it might," she said hopefully. "With such a shortage of men it doesn't seem fair that so many women should be condemned to grow old without even so much as the occasional lick."

"Ha ha ha. But what about all those grannies who still remember how their sons died? They're a powerful vote, you know."

"I suppose," Persephone conceded.

There were indeed grandmothers like Persephone's own, who had taken the family Labrador outside and shot it after her son and grandson caught the fatal canine-carried disease. The dog had been Persephone's birthday present, but Grandma Rayfield had placed a single bullet between the puppy's eyes while it was still wagging its tail.

Persephone refused to visit her dying brother in the hospital after that. After all, it's hard to choose between your dog and your brother at the age of five.

"Anyway, what did you go for?" Serena was trying to steer the conversation back into cheerier waters.

Persephone looked at the printout the agency had given her to take away, then pushed it across the table to Serena. "Pretty similar to yours, really. Blond. Muscular. English."

"Great. They can be playmates. Pity we can't mate them together and start a line of our own," she added.

"Yeah," Persephone laughed. "I suppose we could always mate them with us?"

"What? Are you kidding? And get varicose veins. And haemorrhoids. Not to mention getting as fat as a house."

"Would that bother you?"

"No," Serena lied. "But why risk birth defects when you can get sperm from your friendly gynaecologist that's been more rigorously tested than the air bags in your car? I'm having mine done surrogate and I'm having a girl. Who's going to look after me in my dotage otherwise?"

"I thought you were going to go for euthanasia when things got that bad."

"I know. But right now I'm enjoying life so much. And I have to outlive Peter, don't I? I mean, who'll take on a fifty-something love muffin that's already been trained in all the worst habits? He'd starve without me."

"It's one hell of a commitment, isn't it?" Persephone said gravely.

"Well," sighed Serena. "It's certainly not just for Christmas."

A week later, Persephone's doubts dissolved when she saw her very own man step off the plane from Heathrow. The woman from the agency had called only that morning to let Persephone know a match had been found and would be flying into Los Angeles that very afternoon. The boy had been given special permission to leave Britain to visit his dying cousin (a number of false birth and marriage certificates had linked him to Persephone), and Intermates had to get him on a plane before it was discovered that he had no such American relation. The agent gave Persephone the option to have him lodged in a hotel until she was ready to receive him, but Persephone would hear of no such thing. She wanted to see him straight away.

She recognised him the second he strolled out of Immigration. It seemed as though every head in the arrivals lounge turned in his direction. Persephone had suspected that the agency's faxed picture of him had been enhanced, so she was delighted to discover that in actual fact it was almost true to life. Only his hair wasn't quite so blond. But that wasn't too surprising. Even since they got a decent climate it seemed that the English preferred to stay indoors.

The boy recognised her too, and he instantly switched on a smile that lit up his face and made Persephone feel like she was the only girl in the whole airport. When he was close enough, they shook hands quite formally, but Persephone was sure that she could feel the first tremors of an orgasm at even such a fleeting touch.

"Persephone," he murmured her name as though he wanted to caress her with his voice.

"Jed," she replied, but her voice cracked on the single syllable. Already she wanted him to touch her again. She was trembling when she laid a hand on his bare forearm to guide him in the direction of the car-park. She hadn't touched a man since she'd said goodbye to her father at the Cedars-Sinai Clinic so many years before.

"Did you have a good flight?" she asked banally. She wanted to hear him talk more too. She hadn't got an idea of his accent yet.

"It was bearable," he told her. "But I had to fill out a load of paperwork before they would let me disembark. I thought at one point they were going to send me back. They must have cross-referenced my visa-number with the entire Los Angeles telephone directory."

"But you got the right stamp," Persephone panicked.

"Indefinite leave to remain," he reassured her. "Though I can't think why they're so precious about letting people in here," he added now that they were outside the terminal. "Looking out of the window on the way over, I couldn't see anything but dust."

"Over some of the finest mineral deposits in the world," Persephone laughed. "But I hear it's pretty green where you come from."

"For the rich."

"It's green for the rich here too. I can't wait to show you where we're going to live," she began to babble. "You'll love it. I have a pool. Oh, look, there's my car. The New Generation Ferrari. Picked it up this morning. I haven't got used to it yet. I was still looking for the old Ford. Silly girl!"

Jed smiled appreciatively. "You're a woman of great taste."

Persephone shivered at the compliment.

"You can drive it, if you like." She thrust the keys into his hand. The Ferrari held no real appeal for her, but watching Jed's face as she handed it over was something special. Angela had been right about men and cars.

He settled himself into the driver's seat and adjusted the rear-view mirror. "Pity it's automatic," he flirted. "A gearstick would have given me an excuse to touch your knee." He touched it anyway, and the warmth of his hand flooded Persephone's body. She giggled. It was a strange laugh that had never passed her lips when she flirted with potential female lovers. She could see already what Serena had meant when she said that being with a man was different. Her tongue was tied. She couldn't think of anything intelligent to say. That never happened when she was out with a girl.

"You ready to go home?" he asked her, with a sparkling wink.

"I can't wait," she replied. "We can't get there soon enough for me." However, when Jed started to rev the powerful engine, Persephone managed to restrain her excitement just enough to tell him: "Jed, I'm afraid the speed limit's twelve miles an hour."

It was understood from the start that Jed would move into Persephone's apartment with her and that she would give him an allowance. It was in the terms of the introduction agency's agreement, exactly as she had been warned. But Persephone didn't regret it for one second. Even though Jed's allowance, including the agency's ongoing commission, was double all her other outgoings, she was sure that she had got a bargain. Jed fitted into her life so perfectly, it was as though he had always been there. And the little things that he did! Enchanting. Like the first time she saw him use a spoon to eat dessert. That was so cute. As was the way he pronounced the 'h' in herbal.

Persephone started to work from home much more often than before. Lots of people did, of course, since commuting was hell and, despite the predictions of the technophobes, tele-working still didn't require you to change out of your pijamas. But, living alone as she did, Persephone had always preferred the occasional human interface of the office.

Now she didn't need it. Now she had human contact on a plate. She would wake up every morning and still gasp in surprise when she realised that her hand was lying gently on the warm waist of a real living, breathing creature. When she could finally make time to have a glass of wine with her neglected best friend, Persephone swore to Angela that there were nerve cells in her body that had never been used before Jed stepped off that plane.

Other parts of Persephone too were being activated for the very first time. As Jed was sleeping, she would look at his face and wonder how those people on the other side of the Pacific Ocean could possibly be leaving tiny babies to starve to death when they might one day grow up to be as lovely as Jed. Whereas before she had been pretty much indifferent to the bleating of the television campaigners for human rights, she began to see their point and even sent a cheque for half a million. Surely an economy as great as California's could afford to carry a few more supremely decorative angels like the one she had adopted for herself? They didn't have to take up valuable jobs if they were placed with women who would pay for them. They didn't even have to go to school. Persephone began to fondly daydream that she might one day be rich enough to damn the Gaians and have a son.

A few weeks after Jed arrived in Los Angeles, Serena threw a party. Everyone who was anyone was there. Persephone and Jed drove the short distance to Serena's house in the Ferrari. They wanted to make an entrance—it was the first time they had been out together in public, since immigration laws required that Jed stay within the grounds of his American sponsor's house until he had been vaccinated and checked out by a state doctor. He had been passed the day before the party after an examination that had seemed to take forever, with the doctor lingering, most unnecessarily Persephone thought, over the birthmark on Jed's left buttock. Now, with that trauma firmly behind her, Persephone was particularly looking forward to seeing Angela again. More specifically, to seeing Angela's face take on the pallor of a green-lipped mussel when she saw what she could have got for the money she had spent on a RIMMER.

Angela's solar-powered eco-pod was already parked up outside Serena's house. Serena's boy Peter was standing at the door, ready to relieve the guests of their coats as they arrived. Persephone squeezed Jed's hand tightly as he helped her out of the car. At the door, Peter greeted them excitedly, snatching Jed's hands and then hugging him tightly as though they were long-lost brothers. Persephone remained more composed, though her chest was heaving with anticipation. They walked into the sitting room. People who had been deep in conversation moved apart to let them pass. Surreptitious hands reached out to touch the hair on Jed's bare arms, the shirt-covered muscles of his back.

Serena and Angela were standing by the pool. Serena turned and widened her eyes in delight.

"Isn't he just perfect!" she purred as she ruffled Jed's golden mop. Jed stared straight past her to where Angela stood nursing a coke.

"Hello, Jed," she said quietly. Persephone couldn't help feeling pleased at Angela's response. She had even blushed. Though Angela would never admit it, she was obviously impressed.

"Oh," Serena was squawking. "You must bring him over here, to the shade. I've got some beer in specially. I know how much they like it. And later on he and Peter can go off and play together. I'm sure they'll get on so well. He has had his inoculations now, hasn't he?"

Jed was pushed down into a seat deep with cushions. Persephone and Serena took up positions on either side of him and talked across his well-muscled chest. Persephone had her hand firmly on his left thigh. Serena contented herself with stroking his still-pale arm.

"Angela doesn't look very happy," Persephone observed after a while, when Jed had wriggled away to procure some drinks.

"Mmm," laughed Serena. "Neither would I with nothing but a RIMMER for company. She's split from that Jennifer girl and it's put her in a foul mood. I told her she should get herself a nice malleable man like we have instead. She could afford it. No temper tantrums once a month, I told her. No rowing because you've both turned up to some party wearing the same dress. She told me that if I thought I'd chosen the trouble-free option, she wanted to be around for my rude awakening."

"What did she mean?"

"She said that even a 'sweet thing' like my Peter would 'revert to type' given half a chance. Said I should keep him away from your Jed if I didn't want any trouble. Something about the pack instinct. Can you believe that?"

Jed had forgotten the drinks. Instead, he and Peter were taking it in turns to jump off the little springboard into the deep end of the pool. Serena and Persephone watched them appreciatively.

"I tell you, Persephone, Angela's lost it. Must have had the voltage on her RIMMER up too high," Serena concluded. "Valeria, can you make sure they're getting enough canapés over by the pool?"

Valeria the house-girl sashayed across the garden with a tray in each hand. Persephone wondered if she was exaggerating the swing of her narrow hips even more than usual as she sauntered past the boys. Peter jumped up to relieve Valeria of one of the trays. Jed followed suit. Which just showed how well-bred he was, thought Persephone.

When Serena next had a pool party, almost two months later, things were very different. Peter stood at the door again to take the coats, but when Jed bounded up to him, Peter welcomed his friend into the house with the enthusiasm of a sloth for a half-marathon. Serena too seemed out of sorts. She was grey around the eyes. Valeria had gone, to be replaced by an older woman who could just about manage one tray at a time.

Persephone greeted Serena with a kiss on each cheek. "Darling, we haven't seen you for ages. What's been going on?"

"You mean Angela hasn't told you?" Serena said dryly.

"I don't see as much of her as I used to," Persephone said. "I've been busy around the house, as I'm sure you can imagine."

"That's all I can do. Imagine."

Angela was sitting by the pool already. She looked different too. Though in her case, there was a marked improvement to her usually pinched-up face. Persephone looked at her questioningly. "I won't say I told you so," Angela began. "But it seems that after we were last here, young Peter got a whiff of the hormones. Serena found him wasting himself on Valeria."

"I've been in therapy ever since," Serena sighed.

"But I thought he was totally faithful to you," Persephone said in surprise.

"You can never be sure that the urge is out of them," Angela said.

"But what did you do?" Persephone asked. "I mean, you got rid of her, obviously. But are you sure you can trust him now?"

Peter and Jed were dangling their feet in the pool. Peter began to peel his shirt off. He was thicker-waisted than Persephone remembered. The fine muscle tone she had envied in him before getting Jed was being dissolved into flabby curves.

"Oh, yes," said Serena bitterly. "I think I can trust him now."

Angela left shortly after that. It was clear that Serena had no more time for her prophecies. With Angela gone, Persephone turned her attention to comforting her hostess. Peter was still almost the man she had been so fond of, wasn't he? And couldn't she almost be thankful for the fact that he had lost his tendency to be a tad aggressive sometimes, along with his joie de vivre?

Jed returned from the poolside, looking a little drained.

"I'm feeling a bit sick," he told Persephone. "I think I'd like to go home." She felt his forehead and offered to go back to the apartment with him, but he wouldn't have that. He said that he didn't want to spoil Serena's party by taking away one of her favourite guests.

Persephone handed over the car keys and promised that she wouldn't cut her partying short on his account, but after a while she found she couldn't concentrate on Serena's meaningless chatter anymore. She punched Jed's number into her phone and waited nervously while it rang and rang at the other end until eventually she was passed on to an answering service.

While the pre-recorded voice asked for her message, Persephone had a sudden vision of Jed lying helpless on the bed, able to hear the phone but not able to reach it. Hadn't the Parvo started with the victims feeling just "a little sick"? He could be gasping his last while Persephone listened to Serena wondering aloud whether a state-of-the-art music system would have been enough to stop Peter from straying.

Persephone made her excuses.

Persephone crept into the house. If Jed hadn't answered her call because he was simply asleep, she didn't want to wake him. But as she took off her shoes so that she could walk silently across the tiled floor, she heard an unearthly howl from the bedroom.

Persephone panicked. Only a week before, one of her business associates had been strangled in her own bath. Fully aware that, despite her hours in the gym, she was hardly in a position to defend Jed or herself against a maniac, Persephone hurried outside again. In the locked shed she found the rifle that her grandmother had used to kill the pet Labrador all those years before.

But the scene which greeted her when she quietly pushed open the bedroom door was worse to Persephone than murder. Jed's naked buttocks humping comically. Angela's small white feet pointing towards the door. Neither heard her step inside. She let them continue until first he came, then Angela, moments later. They were squeaking. Grunting. Animal noises.

Persephone raised the barrel of the gun and put her eye to the sights. She focused on the back of Jed's head and felt her finger tense. But Jed collapsed upon Angela's shivering body in postcoital bliss before Persephone could bring herself to squeeze the trigger. She lowered the gun silently.

"You've got to help me get out of here." Jed had raised himself up on his elbows and looked down at Angela, who stroked his back soothingly in reply. Persephone felt a layer of ice form over her heart as she remembered the first time he had looked down at her like that, as if he held the world in his arms. "I can't stay here any longer, knowing what could happen if I stepped out of line."

"Persephone wouldn't do that to you, Jed," Angela said softly. "Trust me, sweetheart, I've known her for years. She's my—"

"Best friend?" Persephone completed the sentence.

Jed and Angela turned suddenly towards the door, their eyes widening in horror as they realised that the shadow in Persephone's hand was her gun. "I think you can put that in the past tense, Angela."

"No, Persephone. You don't understand..."

The lovers scrambled apart. Angela covered herself as best she could with a discarded shirt, since Jed had snatched up all the bedclothes to defend his genitals.

"All that bullshit about not wanting a man," Persephone spat. "All that earnest advice you gave Serena about the trouble with having a lover who's a slave to his hormones."

"I couldn't help it. I meant everything I said to Serena about men, but Jed was different. I didn't want it to be like this, I swear. It just sort of happened. You're not going to shoot me, are you? Say you're not going to shoot me, Perse?"

Persephone raised the gun again.

"Don't do it, please. Please, Persephone."

It was Jed, pleading from beneath the duvet.

"Beg for me," said Persephone. "Beg."

There wasn't much that Jed could do but agree to dig the hole. Persephone had him out in the garden as soon as darkness fell, on his hands and knees in the dirt, scrabbling away without a shovel. As he was digging, she knew that Jed would be planning his escape. But where would he go? Back to England to face the Gaians and almost certain death?

No, she thought almost sadly. He would stay. He probably wouldn't even have the guts to leave her house.

Persephone had no doubt that there were plenty of women out there in the vast sprawl of Los Angeles who would gladly take Jed in, but once he stepped outside her gate without the safety of the car, it was more likely that some mutant veteran from the African wars would get to him first. She had made sure it was well-known from the day Jed arrived in America that there were at least two of those crazy creatures in the canyon below her house, just lying in wait for a man to pass while they laboured under the bizarre notion that fresh sperm would give them back the humanity they had lost to chemical warfare.

"Come inside, Jed," she commanded, when the hole had been filled and smoothed over. Jed sat in the corner of the vast cool lounge, never taking his eyes off Persephone as she sat in her chair, mobile phone in lap, pondering what to do next.

She had not asked Serena what she'd had done to Peter after finding him with the house-girl. For once, Serena hadn't been bursting to boast about the qualifications of the surgeon or the astronomical cost, as she had done after each of her cosmetic ops. Jed's face seemed to register a brief flicker of fear as Persephone picked up the phone and asked for Serena now. He couldn't know, could he? Sometimes it was as though he were telepathic.

Serena not only gave Persephone the number of her doctor, she offered to drive her there next morning.

Jed sat in the back of the car. His eyes reflected frightened and small in the rear-view mirror.

"Peter looked just like that when I took him in," Serena told Persephone, while they waited at some traffic lights. "It's almost as if they know what's going to happen, isn't it?"

"I feel guilty."

"Why? He'll be liberated from all those terrible urges."

"Yeah. But I won't."

Dr. Stoughton welcomed Persephone into her office and had Jed sit with his back to the door that led onto the surgery. When the papers had been signed and the patient led away, Persephone joined Serena again, and Dr. Stoughton's secretary brought the women coffee and biscuits. Moments later she reappeared with a trolley, carrying a large brown cardboard box which emitted scrabbling sounds.

"Dr. Stoughton thought you might be interested in these," the secretary drawled. "Law was finally repealed yesterday afternoon. You can keep dogs now so long as they're properly checked out."

"And these must be," Serena said, lifting one of the golden fur bundles out of the box and rubbing it against her cheek. "You know, Persephone, Dr. Stoughton specialised in dogs before the plague robbed half her trade."

Persephone blocked her mind to the anguished pleas drifting in from the surgery and replied levelly, "She must be very pleased that things have changed again. Are these really Labradors? Wow. I haven't seen one of these since I was a girl."

The dogs were expensive, but it would be several years before they were no longer a rarity. And the wonderful thing was that while they were still rare, the dogs wouldn't know it and try to take advantage of the situation as Jed and Peter had. Serena bought two. She sighed that she would have to put her new pool on hold for another year to afford this latest luxury, but since his operation Peter hadn't been much good for anything but gardening.

"He can walk them," Serena said. "Might help keep some of that fat off. And you should get one," she told Persephone. "Jed will need the extra exercise too once he's been done."

Persephone chose her new pet from the five that remained in the box and handed over her credit card. Dr. Stoughton emerged from the surgery, peeling off her gloves as she walked. She had a delicate smattering of blood on the front of her white overall. Persephone couldn't help wincing at the thought of the needle going into her poor boy's behind.

"How long is he going to be in recovery?" asked Serena.

"Recovery?" Dr. Stoughton looked surprised. "Ms. Rayfield can take him with her right away if she really wants to, but I thought she might like me to sort things out from here."

Persephone nodded. "That's probably a better idea."

Serena looked confused.

Persephone sniffed back a tear and nuzzled her nose in the tiny Labrador's fur. "I don't believe in letting dumb creatures suffer," she explained to her friend. "I'm afraid poor Jed's had his last walk."

On the slow drive home, Persephone reassured Serena that the dog would go some way to helping her get over the loss of Jed's company. A far greater loss would be Angela's companionship. Persephone still couldn't quite believe that twenty years' friendship could have been lost in a flash of deception and jealousy.

"What did you do to her?" Serena asked insistently. "You're so soft on people who upset you."

"No," said Persephone. "I think I made her sorry." She had a sudden, clear memory of Angela kneeling naked on the bed, hands clasped in prayer. Jed was cowering behind her like a cur, begging for Angela's life and yet using it to shield his own.

"I'll give you anything you want," Angela had pleaded again and again. Persephone's eyes misted over with tears.

"Okay," Serena said. "You don't have to tell me what went on between you guys, if you don't want to. Say, these dogs are just too cute. I think I'm going to call this one Peter Two. How about yours?"

"Jed, I suppose," said Persephone. "It'll save me having to get used to another name about the house."

The traffic had ground to a standstill again. Serena took hold of the wide lapels of her shirt and used them to fan her neck. "Phew. All this heat used to make me desperate for sex when I got home. Aren't you going to miss that?"

Persephone suddenly realised she had never thought of it as sex. She had hoped that she and Jed were "making love." Sex was what he had been doing with Angela that terrible afternoon. The tight buttocks bobbed frantically over the white thighs once more. She had made Jed strip the filthy bed and bury the sheets in the garden. Angela thought those sheets were going to be her shroud.

"I'll give you anything you want," she had begged one last time. "The most valuable thing I own."

Back in the car, Serena continued, "I could have lived on three lots of sex a day. I never even thought about chocolate while Peter was in full working order. Guess I'm going to have to buy one of those virtual reality sex things now. It's not the same, but I'm not going through all that heartache with a man again. No way. You want to go halves on something with me?"

Persephone smiled at last. "No thanks, Serena. You know I'm not too good at sharing."

Besides, Persephone had very good reason to believe that she would soon be receiving a luxury RTI. Low mileage. Fully accessorised. And just one careless prior owner.

A NIGHT ON BARE MOUNTAIN

BY GRAHAM CHARNOCK

The world ends at midnight, Gance. Have you heard?

Yes, it's me—Venn. Still crazy after all these years. Still singing old forgotten songs to myself.

Remember when we played Trivial Pursuit in that brothel in Cairo, next to a command bunker? Scuds were flying overhead. The fundamentalists had broken the precise positioning code on the Axis navsat network and any moment we expected one almighty mother of a portion to be delivered to our pie.

There were four of us. Me and you lot—Cleo, Athene and you, Gance. The three weird sisters, as I called you, who tangled the skeins of weird web worlds together. The mistress seamstresses.

God, what a night that was.

Thumpety thump overhead. Crackety crack in our bones. And the clickety click of the rolling and tumbling dice. We may or may not have been naked, sprawled on silk sheets. Probably not—false memories of those fine times abound—although I seem to remember flesh, bare and white, tracked with reworked scars, moulded over with collagen, new maps on ancient contours. I still had some skin left in those days.

And I remember that as we settled over that stupid board I threw a six and moved onto movies.

Question to Cleo, first on the right:

"Which novel about vampirism was filmed twice, first starring Vincent Price, and later Charlton Heston?"

Cleo giggled and drew her pretty dimpled knees together. She didn't have a clue. Poor clueless Cleo, but God she was sexy. I was hasty, intemperate, wired, as always, passing over sex to try to get at the truth.

Quick, Cleo!

You've got to be quick.

Snap! Really quick.

Bop!

Stab and jab. Float like a butterfly but sting like a bee.

Come on! Quick, girl!

No, you're too slow, you're dead already. You've got to think faster than a speeding bullet if you want to stay in this game. Think on your feet. No, think on your keypad, think on the chill platinum tags of your neural implant, that thing at the back of your ear you scratch with the same degree of neurotic affectation as a soppy schoolgirl pulling at the tags of her untidy hair. Think on the slim needles of steel you keep curled beneath your fingertips, grafted there by the good Dr Whatsisname. If you're not quick you're dead, and your name will be legion. You will be one of the fallen angels and not a Cairo Baby. No, nevermore.

I was trying to give her a clue, but she took it the wrong way and burst out crying. The rest of us cuddled her. Gance, you scolded me for being stupid. We kissed, I remember. Athenc poured herself a drink and stood apart from us, softly scouring her nails against the breeze-blocks that formed the wall, as if she were sharpening them up in anticipation of some act of vengeance.

The Scuds poured down. Any one of them could have wiped us out, in the middle of our inconsequential game, with so many questions left unanswered. What a way to die. I thought at the time there must be a better one, but it's taken me this long to figure it out.

Gance, these are our good days but they are running out. Our history lesson for today is on how history lessens, and I'm sending out this message to you, Gance, because I'm alone, and I need to know if this is really how it ends, with all of us separated forever.

I don't mean this conjunction of the planets they're all talking about, which is supposed to pull out all the nails that hold the continuum together but has so far simply pulled every prophet of doom out of the woodwork. All the planets and their satellites in line for the first time in thirty million years and that big comet, Alcephus, steaming through the middle of it all precisely on the stroke of twelve. Rubbish, Gance. I can't rely on some shoddy piece of cheap cosmic clockwork to do the job for me.

Are you still my Cairo Baby? God, I hope so—I could use your needles now. My body is in its last passage towards something spectacular. My body is a comet, its outer skin burning off as it Tuttles back to Earth after its latest trip around the cosmos. I burn, baby. I am Alcephus. I need your cool relief.

Here's a tale. There was this bee buzzing around my virtual rose. He was slow and I zapped him. I'm not too well-coordinated these days, but a friend of mine made me this headset and all I had to do was roll my one remaining eye in its socket and that did it. Trouble was the virtual hand of God only caught him a glancing blow that half crushed him—his abdomen was split and squashed, it was hanging off his thorax by the slimmest thread, a strand of neural tissue connecting his brain to his sting, his nose to his arse. He still had some kind of information flow, triggering and firing his crushed synapses, but he was damaged. Still he buzzed around the rose. He didn't even know it was mid-winter in the virtual latitude I'd constructed for him and that the rose was just a dead calyx, a memory of summer. He just didn't know he was out of the game, babe, but he carried on.

I know we're all damaged, one way or another, Gance. But we have to carry on. Don't we?

Which novel about vampirism?

None of you knew the answer years ago in that bunker. I'll run it past you one more time, and here's a clue, think God, the Bible. "My name is legion: for we are many." Gospel according to St. Mark, 5:9.

I can imagine you scratching you head, Gance, just like then. Maybe you think I've found religion in these final terminal days as the old clock clicks over? It's like we're riding in Nostradamus's taxi, and God only knows if we'll be able to find the fare at the end of the ride.

But there is no God, babe, unless it's me.

Time's up.

I know the answer still eludes you.

The big gong has just gonged. The audience is going Aaaaahhhh with that sad dying fall as the chick in the net stockings and high heels takes your arm to lead you into the wings, simpering that stupid smile at you all the while with those gleaming teeth.

Why did God give primates an opposable thumb and perfect teeth? So they could pick and grin.

You're a loser, babe, so why don't I kill you?

I Am Legend, for God's sake, by Richard Matheson! They filmed it first as *The Last Man On Earth*. That was with Vincent Price, and then they did it again in 1971 with Charlton Heston, but they called it *The Omega Man*. It's one of the great ideas—vampirism is nothing supernatural, it's a disease that's all, you know, like TB, or CJD, or Ebola or Michelangelo. Oh yes, very like Michelangelo because it's ubiquitous; they're a new different breed of humanity and everybody's joined the club. Except for Charlton Heston, who's immune. Poor sod. Poor silly sod.

You owe me a forfeit from that night years ago, you and your sisters, and I'm calling it in. I'm throwing a party. I don't anticipate it will go on much after midnight.

There's a small, select guest list. Just me and you, Gance, and your sisters, like old times.

Have you seen your sisters lately? The last time I saw Athene was when I fucked her brains out in Amsterdam.

I came across her table-dancing in a sushi bar on Engheitstraat, a dive underneath some damn bridge, half in and half out of the *Oude Zijds Voorburgwal*. There were condoms floating in the grey water; the stuff they served up in the bar looked and tasted like goddamn condoms.

Athene was far gone that night. She didn't even recognise me. Her head was full of weirdness. She thought I was just another punter. She went on and on about how she could cure pain and bring healing, thanks to Dr Whatsisname, who'd grafted needles underneath her cuticles directly connected with the meridians of her yin and yang. She was a walking acupuncture babe, she said. As if I didn't know.

Athene was very gabby that night. She told me she was a big bright green pleasure machine. Then her eyes went glassy and she told me twice how she'd once met Stephen Hawking in a bus station in West London. That's how far gone she was.

I took her back to the American Hotel, strapped her to the bed and fucked her stupid for a week. She still didn't know who I was. She said I was sick, if I just untied her, she could help me with those bloody needles of hers, fuse the meridians of her yin and yang with mine. She showed me one, flicked it out from her curled forefinger like a cartoon cat mesmerizing its victim with its claws.

But I wasn't having any of it, not that night. I saw vengeance in her eyes, Gance. Maybe she recognised me at last and remembered that night in the Cairo brothel. I wasn't convinced she had my well-being at heart.

I wonder how she got off that bed anyway, after I'd gone?

I'm really looking forward to seeing you again, Gance. It will be like old times. We can Frug together, do the Twist, the Madison— anything you like. And there's so much to talk about.

You heard how Elvis landed on the moon, right? One small step for a man, one giant leap for King Creole.

And how about when the Mars probe found that time-machine and they used it to go back and kidnap Phil Dick? That was some stunt, eh? Was he gob-smacked to see what a cult hero he'd become?

And you remember when Bill Gates stepped into the world's first artificially generated Wormhole in Austin and stepped out in Lubbock? And you remember the horrible shirt he was wearing?

GRAHAM CHARNOCK

Gance, you do remember recent history, don't you? I would hate to think I was investing all this time and stuff in someone whose brains had been fried by too much prion-protein.

Come on, climb into my web, and toast eternity with me as time tick ticks out. Let's have one last dance together.

Bring a bottle.

Gance—one important thing—you're wondering why the guy who handed you this package exploded? He had an implant of explosive gel in his left buttock and instructions to self-destruct within two minutes of handing it to you. Sorry about the mess. He was only a mule. Don't feel bad about it. He knew what he was getting into, and, anyway, he believed in Alcephus. His dependents have been very richly rewarded.

I'm a wealthy man now, Gance. It's just that I choose to live this way.

Haven't you been following my career? Shame on you. One good idea, you know? Wasn't that what we always said? One good idea and tomorrow can take care of itself. And I have a million good ideas now, all primed and ready to scuttle out across the universe. A million Silver Beetles hurtling down their very own Helter Skelter.

You never believed in me, Gance. Of the three of you, you were the only one I could never get to because basically you thought I was shit.

But check out the package.

Just pictures, third-generation Polaroids produced on vintage stock. There's no test known to man can prove them real or otherwise. Smart eh? The sort of thing a shit might think of.

Recognise your sisters? I know it must be hard. How many years, after all? Yeah, that's Cleo, retrieved from the memory bank of a photo-booth in a supermarket in Canton. She's so thin and frail, isn't she? What can she have been doing with her life? And this one's Athene. Someone took her photo on a digital camera as she waited to withdraw money from a bank in São Paulo. Just the face again, but she's touching her cheek with her hand. Where did she get those scars, do you think? It looks as if someone beat her up.

They both look as if they've been through hell, don't they, Gance? But never mind, their suffering is nearly over. They're here with me now, Gance. Like the rest of us they'll die at midnight. This will be your last chance to say goodbye to them.

I know you won't let us all down.

"How do you kill a lobster?" Thumb asked Gance.

"I don't know. Throw it in a pot of boiling water?"

"No. If you look at the back of the head where the carapace meets the thorax, there's a small natural cross mark. Stab a knife in there and it severs the spinal cord and the thing dies instantly. It's the end of the world as your crustacean knows it. Now the question we have to ask is, when it's so easy to kill the lobster without causing unnecessary pain and suffering, why do most people just toss it in a pot of boiling water and watch it thrash itself through a painful death?"

"You tell me."

"I'm still trying to work it out, darling, but I think it's something to do with the human condition. Where's your ride, anyway? I'm getting cold."

"Here it is now," said Gance, pulling a black leather glove onto her left hand.

"It's gone dark," Thumb said.

"Just keep quiet. Jesus doesn't know about you."

Bats clattered behind the broken windows of the shopping mall's abandoned atrium. The creatures congregated in void spaces of the mall's roof, where the slow leak of CFCs from the stacked piles of abandoned freezers and refrigerators collected and generated an area of warmth.

Their high-frequency chattering made her cuticles tingle.

Gance stepped out from the shadow of the shabby colonnade as the cab approached. Its riding lights were a dim glow. A clammy wind was blowing in from the east. Somewhere over the European Shelf it had picked up a freight of macro-spores, and now it was shedding them in a localised fall. The flakes spun off across the empty car-park, deliquescing into slush-pools where they landed.

The cab rolled to a halt and Jesus Hitler wound down the window an inch to check it was Gance. Gance scowled at him, and punched at the door handle.

She climbed in beside him, brushing flakes out of her hair.

"What's the score, Gance?" Jesus asked.

The engine stalled as he tried to move away. He twisted the key three times in the ignition. The starter motor whined and groaned and then the engine finally fired.

Jesus pulled a cigarette out of the pack on the dash and offered it to Gance, but she brushed his hand aside brusquely. He took one himself and tried to light it from a plastic disposable lighter. Nothing happened. He threw the lighter aside in disgust.

"Fucking move," she said. "I'm late enough already."

Jesus shrugged. "Take it easy. Where are we going?"

"I told you."

"The levee? You're serious? Your last night on Earth and you want to go to the levee?"

"Don't tell me you've bought into that paranoid shit, Jesus. A good Russian Orthodox Jew like you?"

"It's true. The world ends at midnight tonight. Everything is all lined up and Alcephus is on its way. You know why that big ball of rock is called Alcephus? It's from the Arabic—*al cifr*. The big zero. That's what's waiting for all of us, babe."

Gance shook her head tiredly. "We had all this crap before with the Billennium," she said. "We're still here."

"Yeah, but this time even the Pope is saying kaddish."

"Look, if it's Armageddon tonight, why aren't you curled up somewhere frying your brains?"

"You know how it is—gotta make a living somehow. Why ain't you?"

"Same reason."

They drove across the city. The vast edifice of the levee loomed up ahead. Cityside, it was a blur of animated neon and coruscating advertising displays.

From a third of a mile above, Gance heard the deep-throated doppler-descending roar of an incoming HTOL freighter, then the abrupt silence as its ram-jets cut out and it hit the runway strip; the noise was replaced with a low subsonic rumbling which she felt deep in her bones.

Then suddenly they were plunged into darkness as they arrowed into the narrow eastbound tunnel that transected the massive earthwork.

They emerged from the tunnel into a darker, shadowy environment. The eastern face of the levee was dark in contrast with the city-side. Feeble lights flickered across it like evanescent transient demons.

They entered the network of causeways that threaded through the lagoons of the tidal basin. Gance directed Jesus towards the arm of the levee that curved away to the south. Jesus's vehicle lurched and faltered at every slight gradient it encountered. The night sky was thick with clouds, obscuring any view of the web far above.

There was a small cut-out fir-tree shape made of cardboard dangling from the cab's rear-view mirror. It smelt faintly of an odour Gance could not place. She flicked it with her finger.

"Where did you get this, Jesus?"

"A guy I know makes them."

"What's the smell?"

"Does it smell?" Jesus said. "I never noticed."

Gance took the two Polaroids out of her top pocket and showed them to Jesus Hitler.

He studied them, steering with one hand as they rattled across a rough pontoon that straddled one of the many creeks and inlets.

"Cute girls," he said. "Who are they?"

"They're my sisters."

Jesus squinted sideways at her. "I can't see no family resemblance," he said.

"Sisters under the skin," said Gance. "The guy I'm going to see tonight claims he's holding them. But they've been dead a long time. They both got caught in that Sarin incident in Geneva twelve years ago. What do you make of that?"

"The guy is crazy."

"That's what I figure. And it gives me a bad feeling."

She took the Polaroids back from him and carefully put them back into her pocket.

A NIGHT ON BARE MOUNTAIN

Jesus braked suddenly. Up ahead was a crude counter-weighted wooden barrier painted with black and yellow slashes. Off to one side a group of men were gathered round a brazier. One of them approached the cab. He held a long skewer in his hand, with something pungent and smoking impaled upon it. Jesus wound down his window a few inches.

"We're having a little Last Supper," the man said, "Want to take communion with us? Care to partake of the flesh of the prophet? A man shouldn't seek redemption on an empty stomach."

"Fuck," said Jesus, shooting a glance at Gance. "What now?"

"Give him money," Gance said.

Jesus passed over some coins and the man handed through the skewer of cooked meat in return, then stepped back to raise the barrier for them.

They drove through. When they were well clear Jesus hurled the kebab into the lagoon. It spat and sizzled as it hit the water and immediately a dark shape rose to the surface in a flurry of bubbles to snap at the meat.

Eventually they drove onto the rubble-strewn beach that ran the length of the levee's southernmost extremity. Jesus drove slowly here, threading his way through the shantytown construction of shacks and crude shelters, trying to avoid the children and dogs. By mistake he turned into a blind alley; as he reversed out, a dog ran across the rubble-strewn single-lane track. Jesus slammed on the brakes too late to avoid it. It disappeared under the front wheels and the cab jerked to a halt, doing a sideways slither as its engine seized, and slamming up against a shack constructed of something that looked like animal hide cured in resin.

Jesus stomped out of the cab and hiked up the bonnet, waving away the scent of burnt rubber and smashed dog.

"Broken fan belt," he reported. "I guess this is as far as I take you. Do you want me to wait?"

"Where would you be going?" Gance asked, as she walked off towards the levee.

"All I need is an old pair of pantyhose," Jesus called after her. "You think I can't find something like that on the night the world ends?"

It started to rain. The microbacterial organisms in the rain congealed almost instantaneously into a clinging film as soon as they struck her PVC jacket.

She shivered, although it was not cold. She caught a twinge from the chewed-up socket in her left armpit. The next big job she got she was going to have the thing ripped out. It was useless anyway—the protocol had been obsolete for a long time. The last time she'd tried to jack directly into a client she'd nearly fried his temporal lobe. She'd left him twitching on his futon, but not before wiping a large proportion of his collateral into an untraceable account.

There were lights up above her on the levee, but they were fitful things, too dim for her to make out anything. She took off her glove so Thumb could see.

Thumb spread a field of random nanolaser, resampled the reflections and shot it up through her optic nerve into the small area of retinal display she had put aside as an overlay. She saw a pool of muddy scum ahead of her she might otherwise have stepped in.

Thumb read the spectral analysis. "Mutated algal froth," he said. "Cloned from something an early Mars probe brought back. Tiny single-celled plants from hell. They can strip your epidermis off within seconds of making contact."

"Nice," Gance said. "There was none of that stuff last time I was here."

"It's a changing world," Thumb said, then, "Something coming, left field. Something fast. Duck!"

Gance ducked. A fat shape shot past her left shoulder.

"Leaping lizards," said Thumb with a chortle. "Rat, actually. Jesus, look at the mother."

Gance did. It was large and feral and it had wings.

"Bat-Rat," Thumb amended. "Scramblegene job. Someone's been playing with an outlawed morphing programme. Something else coming, now. Left field again. Slow this time. But armed and dangerous. Duck!"

Gance ducked. This time it was a narrow-spectrum spurt of laser-blast that overshot her. Thumb fed her the overlay, so she didn't even have to turn her head, not that she had time. The laser painted the rat's hide with a poisonous tattoo. The rat sizzled and somersaulted onto a small mountain of discarded hard-core where it fell apart, weeping into the rubble like slurry in the rain.

Thumb zeroed a grid on the laser source, and built a shape around the wire-frame that quickly took on visual flesh, betraying a slight figure holding a crude laser-rifle. The man was webbed in a dark mesh of containing clothing, and he looked corroded. He hobbled through the wash of dirt and chemical mire, crippled.

"Looks like your date," said Thumb.

The initial damage had been done during the Winter War, in the dirty years after Etna had blown six million cubic metres of itself into the upper atmosphere. The military were big employers in those days;

they recruited Venn the day he graduated from Huddersfield with a double first in nanotechnics, and sent him out to the desert.

He'd only been out there a month when it happened. He was in a group of six techs scheduled for basic combat training, but an order got scrambled and they were sent in the place of some boots, to test a new stinger-field laced with a prototype toxin.

It took them sixteen months to rebuild his sciatic nerve. That was when Gance first met him. She was still in Cairo then, working out of a military brothel on the Shari El Corniche, with Cleo and Athene. After every operation they were called in to clear and rechannel his meridians.

It was dirty work and sometimes the interchange and foldback made her feel as damaged as him, but she'd got to know his nervous system pretty well. She'd come away with maps and templates in her head that she would use later to refine many areas of her technique.

Venn came out on a disability pension and put in a civil suit against the military. While that was in process, he went to the Shelf and fell in with some Palermo suits who ran protection for the freight-loaders that worked the HTOL strip. It was boom time, the geo-stationary web was slowly growing towards total-knit, and all sorts of franchises were taking out leases, from one-man clone-labs to full-scale factory-farm developmental units. The freight-loaders carried rich pickings in raw materials and fabricated plants.

One day he was riding shotgun on a ten-wheeler artic as one of a two-man security team, with a pumper called Clive whose cognitive processes had long since seceded from his forebrain. Clive regularly popped at least two tendons a week. He'd had an ice-job to try to relax his skin but that had broken down and now he looked, walked and talked like a mummy. The artic was carrying a third-rate protein called oca which originated in Peru. It had mild narcotic properties and could be refined in zero-g conditions into something which slowed both the metabolism and the rate of calcium loss; it had become a staple of the web-workers' diet.

The artic was hit by a bunch of kids who blew it off the ramp with a homemade bazooka. The pilot ended up a fine mash of pulped bone and flesh all over his own tachometer. Clive's femoral artery was sliced and he bled to death where he fell. Venn managed to crawl away, but the kids caught him and torched him. They were pretty pissed that they'd bagged nothing more than a load of Peruvian pond-weed.

When the judgement in the civil suit came in, most of it went to pay for Venn's skin-job.

He didn't appear to need Gance or any of them after that; at least, he left them alone.

She heard he'd gone to ground in the levee, bought some second-hand gear—a cryo-vat, and a resonance-field moulder—and set up a molecular-diffusion lab. She'd heard all sorts of rumours, that he was producing functional nanosurgical devices—arterial scrubbers and phage-hunters—that were cheap but out-performed anything from Montreal. Then she'd heard nothing until a mule—a kid with a goatee beard—had come across her while she was shopping, handed her a package, whispered a zen blessing, adopted the lotus position and blown his arse across three aisles.

The rain turned to sleet. They started to climb the levee. She followed Venn. He moved quickly, despite his hobbled gait, and she scrambled behind him. They climbed ramps, traversed ledges, squeezed through barely passable fissures, and leapt from one ramshackle platform to another as they ascended. The path forked and branched endlessly and occasionally she'd find her fingers settling upon a chiselled handhold or a metal rung. The darkness of the levee closed in upon them. Her concentration was centred on following him, almost step-for-step.

Her knuckles and elbows soon became scuffed and scratched. Her thighs and leg muscles started to ache with the exertion, but Venn wouldn't let her pause. He grasped and pulled at her jacket, tugging her on. She surrendered to his manhandling, moving almost blindly, placing her hands and feet wherever Venn directed. Eventually they moved onto a wide, flat ledge, and Venn released her. She looked down, startled at how far they had climbed. There was a peculiar radiance about the lagoons far below them that had not been apparent as she had driven across them with Jesus; they swirled with an eerie, patchy phosphorescence. She looked up: the clouds that obscured the web hundreds of miles above her were as impenetrable as ever.

When she turned around to look for Venn she could not see him at first, and then made him out, signalling to her from a cleft in the levee's face. She followed, squeezing through a narrow opening into an area of pitch darkness.

Thumb painted in the details for her. The walls of the chamber were constructed from crudely assembled breeze-blocks held together by an expanded foam sealant that oozed and dripped in untidy hematic globules and stalactites. There were no true horizontals or verticals evident in the jointing, no right-angles, nothing flush or finished. There was crap all over the floor—straggling fronds of dried and withered vegetation, shredded paper pulped in places by dripping or wind-borne water to the consistency of a primal papier mâché. There were layers of bubble-film and carpet cut-offs, wood chippings and other nameless detritus. It looked to Gance like an eyrie only recently vacated by some huge, disorganised and repulsive bird.

There was an irregular frame of timber on one wall, fashioned in the shape of an old, weathered, splintered wooden pallet. Venn moved towards it, released catches and tilted the construction smoothly upwards to reveal a way through. He brusquely pushed Gance towards the opening.

The space beyond was larger, cavernous by comparison.

The transition as she entered was startling; it was like moving from a dungeon into a bizarrely furnished, sumptuous rococo saloon.

The walls were hung with tapestries. She'd seen something like them before—the new immigrant cultures on the Pacific Rim specialised in them. They were formed not from strands of silk but from the filament-like forms of microscopic thread-worms. These were Methuselahs among their kind, condemned to pirouette eternally on fields of nutritious culture-media bonded to time-degrading micro-laminations. Their slow random dance threw up gaudy patterns, rich with geometrical tribal motifs, bright and hard-edged, like the chevrons of honours carved into warriors' faces.

The rococo element was echoed in the room's other furnishings—an assortment of classic items of furniture—here an oak chest, decorated with crude relief carvings, gargoyles and griffins in miniature, there a military chest with brass bindings, and a Hoshashi table from eighteenth-century Japan, low and lacquered, inlaid with filigree pattern in walnut representing storks in flight, wheeling above a restless ocean which threw up curling wisps of waves.

"Is this stuff for real?" Gance asked.

"Some of it is. Some is reproduction. There are savants who live on the levee who make their living that way."

Venn picked up a cigarette and a matchbox from the lacquered table and handed them to her. The box was constructed from flimsy panels of real wood. There were two abrasive strips on the narrowest sides, and two pictorial panels on the widest—Champion Matches, they were called. The illustration was of a muscled man in a gaudy-coloured leotard. He was bald-headed and had a huge fiery moustache that spread from his upper lip out over both cheeks.

"It's genuine," said Venn. "It's one of the most valuable things I own. The matchbox, that is. The matches are reconstructed. Do something for me—light up the cigarette."

"I don't smoke."

He closed his eyes. "You don't have to smoke it. Just light up the cigarette."

Thumb scanned the cigarette and painted up a message on her overlay: "It's genuine—nothing more toxic than tar and nicotine. Just one won't kill you."

She lit up the cigarette. Venn kept his eyes closed, only opening them when she had finished. He took it from her and stubbed it out.

"Thank you," Venn said. "The most erotic sound in the world—a woman lighting a cigarette from a box of matches. The rattle of the matches in the closed box. The sound of the drawer sliding open. The scrape of the match itself and then the flare. The first inhalation —the air in the nostrils. The moist kiss of the lips separating, and the long sigh of the first exhalation. I just wanted to hear it one last time.

"I wasn't sure that you'd come—this being the end of the world, the night we all die. Then I thought, Gance of all people won't believe. Gance never believed a thing in her life. Gance has cynicism deep in her whore's heart. Gance will come."

"I came because you sounded desperate. What kind of mess are you mixed up in now, Venn?"

"You have an organic chip. Is it smart?"

Gance wasn't surprised that Venn knew about Thumb. He had to have some kind of scanning implant himself to pick them up so easily on the levee.

"Probably not as smart as it thinks it is," Gance said. "I picked it up cheap."

"Ask it who's working on nanoviruses and how close they are to pulling it off."

Thumb said, "Microsoft says they had it cracked ten years ago and they're just waiting for the market to catch up. Bullshit, of course. Stuttgart has the best biotech knowledge base and it says the technology increase required is of an exponential order of magnitude. You can't see the end of the curve. The simplest working nanovirus is that far off. They'll develop a faster-than-light drive sooner. But there is big military funding in there—they might be deliberately muddying the water. The biggest purely commercial concern working in that field is PolyQ. They hold the original Buckyball-Q patents, of course, and now the geostationary web is running hot and generating revenue, they're ploughing a lot of it into nanology and they say real soon now, but they may just be talking it up for their investors. The main money is on a Buddhist enclave in Hong Kong—people reckon when you get that close to playing God it helps to know a little zen."

"I've done it," Venn said. "You're holding it in your hand."

Gance looked again at the box of matches. So did Thumb. "There's something under the label," Thumb said. "A metallic film of some sort. Chips, lots of chips. I don't recognise any of the signatures. It's weird, home-cultured stuff."

"That's just the template," said Venn. He tapped his chest. "As for the prototype, you have to look really deep, in here."

He gave a small gasp and a look of pain passed over his face.

"So much suffering in the world," he said. "It would be a kindness to put the world out of its misery, wouldn't it, Gance, to undo the work of God's sick mind?"

"Is that what this is all about? You've got a virus that can do that?"

"Can't stop it now," Venn said. "My name is Legion; for we are many."

GRAHAM CHARNOCK

Gance shrugged.

"Then we might as well fuck."

Gance undressed him and ran her hands over his wounds. He was in a bad way. The colloid tissue on his old scars was regranulating, flowing, seeping, seething with a barely contained gangrene. She wondered what drugs he had been using to try to control that, and where he had got them from. Some backyard factory in the levee itself, probably. God knows what poisons he'd ingested.

Venn sighed. "I'm looking forward to this," he said. "It's going to be just like old times."

Gance probed with her fingertips, skimming the suppurating flesh. She could sense the damaged meridians beneath the skin, the skeins of power, diminished but still potent, which glowed around the sinews, nerves and muscles in his extended limbs, channeling down into the lumbar regions and the base of the spine. She sensed their alignments and fed on them. She felt the intenseness of an unscratchable itch, the beauty of an unrealised orgasm, as the needles beneath her cuticles spasmed and began a slow extension of their own accord and tracked over the crippled body, searching for a point of access. She called up the maps she had committed to memory all those years ago.

"Make it hurt, Gance," Venn said. "When I hurt, there's only me and then I'm big, bigger than the universe. Then I am the universe. Alcephus be damned. I am infinity. De Quincey knew what it was about. He had an opium dream once—he was on the Brocken, the highest peak of the Harz mountain range. There's an optical illusion up there they call the Spectre of the Brocken, where you see yourself writ large upon the distant mists. We're all something projected. Our ghosts are larger than our selves."

"Be quiet now," said Gance. "I'll help you."

She found it then—a birthmark, a tiny discolouration of skin the shape of a small star on the nape of Venn's neck. That was where she inserted the first needle.

Thumb guided her down from the levee. She stumbled once, gashing the palm of her hand against a skewed rung on an iron ladder. She licked away the blood.

"You're a compassionate woman, Gance," said Thumb. "I'm beginning to learn more about the human condition every day."

"Nuts," said Gance. "You think I killed him? I could have. I wanted to. But I didn't have to. There was something in there, something dark, waiting for me to direct it, to channel it. His nanovirus maybe. It took my energy and just used it to complete a circuit. He killed himself."

"Was he capable of constructing a nanovirus, then?"

"Maybe—but one designed to bring an end to everything? Your bet is as good as mine. I'm not sure I want to know."

Gance looked about her. There was nothing she recognised. Then she heard the throttled-up roar of another HTOL ramming off the strip. She took a step forward and stopped. A dead dog lay in a puddle at her feet. At least she thought it was a dog. It could have been anything.

A burst of sparks shot up into the sky a short way away. Fireworks. Somebody was celebrating something. She heard the sound of laughter. A woman's voice struck up a song and she was joined by a deeper male baritone. She recognised Jesus Hitler's voice. He sounded drunk.

"What's the time?" Gance asked Thumb.

"Past midnight."

"There you are, then. Venn was a fantasist and he got what he wanted. It was the end of the world tonight. For him, and a few others, I'd guess. But we're still alive. For us the world goes on. For the time being. God knows how many other solipsist nuts are out there with grand designs for the extinction of their private universes—which just might include ours too."

"You still have the matchbox," said Thumb. "We can have it analysed. Then we'd know about the virus for sure."

"No—we'll give it to Jesus. He could use a light."

Gance looked up at the sky. The dark clouds that obscured the geostationary web were clearing. One by one the stars were coming on.

LONDON BONE

BY MICHAEL MOORCOCK

For Ronnie Scott

ONE

My name is Raymond Gold and I'm a well-known dealer. I was born too many years ago in Upper Street, Islington. Everybody reckons me in the London markets and I have a good reputation in Manchester and the provinces. I have bought and sold, been the middleman, an agent, an art representative, a professional mentor, a tour guide, a spiritual bridge-builder. These days I call myself a cultural speculator.

But, you won't like it, the more familiar word for my profession, as I practised it until recently, is *scalper*. This kind of language is just another way of isolating the small businessman and making what he does seem sleazy while the stockbroker dealing in millions is supposed to be legitimate. But I don't need to convince anyone today that there's no sodding justice.

'Scalping' is risky. What you do is invest in tickets on spec and hope to make a timely sale when the market for them hits zenith. Any kind of ticket, really, but mostly shows. I've never seen anything offensive about getting the maximum possible profit out of an American matron with more money than sense who's anxious to report home with the right items ticked off the *beento* list. We've all seen them rushing about in their overpriced limos and mini-buses, pretending to be individuals: **Thursday: Changing-of-the-Guard, Harrods, Planet Hollywood, Royal Academy, Tea-At-the-Ritz, Cats.** It's a sort of tribal dance they are all compelled to perform. If they don't perform it, they feel inadequate. **Saturday: Tower of London, Bucket of Blood, Jack-the-Ripper talk, Sherlock Holmes Pub, Sherlock Holmes tour, Madame Tussaud's, Covent Garden Cream Tea, Dogs.** These are people so traumatized by contact with strangers that their only security lies in these rituals, these well-blazed trails and familiar chants. It's my job to smooth their paths, to make them exclaim how pretty and wonderful and elegant and *magical* it all is. The street people aren't a problem. They're just so many charming Dick Van Dykes.

Americans need bullshit the way koala bears need eucalyptus leaves. They've become totally addicted to it. They get so much of it back home that they can't survive without it. It's your duty to help them get their regular fixes while they travel. And when they make it back after three weeks on alien shores, their friends, of course, are always glad of some foreign bullshit for a change.

Even if you sell a show ticket to a real enthusiast, who has already been forty-nine times and is so familiar to the cast they see him in the street and think he's a relative, who are you hurting? Andros Loud Website, Lady Hatchet's loyal laureate, who achieved rank and wealth by celebrating the lighter side of the moral vacuum? He would surely applaud my enterprise in the buccaneering spirit of the free market. Venture capitalism at its bravest. Well, he'd applaud me if he had time these days from his railings against fate, his horrible understanding of the true nature of his coming obscurity. But that's partly what my story's about.

I have to say in my own favour that I'm not merely a speculator or, if you like, exploiter. I'm also a patron. For many years, not just recently, a niagara of dosh has flowed out of my pocket and into the real arts faster than a cat up a Frenchman. Whole orchestras and famous soloists have been brought to the Wigmore Hall on the money they get from me. But I couldn't have afforded this if it wasn't for the definitely iffy *Miss Saigon* (a triumph of well-oiled machinery over dodgy morality) or the unbelievably decrepit *Good Rockin' Tonite* (in which the living dead jive in the aisles), nor, of course, that first great theatrical triumph of the new millennium, *Schindler: The Musical*. Make 'em weep, Uncle Walt!

So who is helping most to support the arts? You, me, the lottery?

I had another reputation, of course, which some saw as a second profession. I was one of the last great London characters. I was always on late-night telly lit from below and Iain Sinclair couldn't write a paragraph without dropping my name at least once. I'm a quintessential Londoner, I am. I'm a Cockney gentleman.

I read Israel Zangwill and Gerald Kersh and Alexander Barron. I can tell you the best books of Pett Ridge and Arthur Morrison. I know Pratface Charlie, Driff and Martin Stone, Bernie Michaud and the even more legendary Gerry and Pat Goldstein. They're all historians, archeologists, revenants. There isn't another culture-dealer in London, oldster or child, who doesn't at some time come to me for an opinion. Even now, when I'm as popular as a pig at a Putney wedding and people hold their noses and dive into traffic rather than have to say hello to me, they still need me for that.

I've known all the famous Londoners or known someone else who did. I can tell stories of long-dead gangsters who made the Krays seem like Amnesty International. Bare-knuckle boxing. Fighting the fascists in the East End. Gun-battles with the police all over Stepney in the 1900s. The terrifying girl gangsters of Whitechapel. Barricading the Old Bill in his own barracks down in Notting Dale.

I can tell you where all the music halls were and what was sung in them. And why. I can tell Marie Lloyd stories and Max Miller stories that are fresh and sharp and bawdy as they day they happened, because their wit and experience came out of the market streets of London. The same streets. The same markets. The same family names. London *is* markets. Markets are London.

I'm a Londoner through and through. I know Mr Gog personally. I know Ma Gog even more personally. During the day I can walk anywhere from Bow to Bayswater faster than any taxi. I love the markets. Brick Lane. Church Street. Portobello. You won't find me on a bike with my bum in the air on a winter's afternoon. I walk or drive. Nothing in between. I wear a camel-hair in winter and a Barraclough's in summer. You know what would happen to a coat like that on a bike.

I love the theatre. I like modern dance, very good movies and ambitious international contemporary music. I like poetry, prose, painting and the decorative arts. I like the lot, the very best that London's got, the whole bloody casserole. I gobble it all up and bang on my bowl for more. Let timid greenbelters creep in at weekends and sink themselves in the West End's familiar deodorised shit if they want to. That's not my city. That's a tourist set. It's what I live off. What all of us show-people live off. It's the old, familiar circus. The big rotate.

We're selling what everybody recognises. What makes them feel safe and certain and sure of every single moment in the city. Nothing to worry about in jolly old London. We sell charm and colour by the yard. Whole word factories turn out new rhyming slang and saucy street characters are trained on council grants. Don't frighten the

MICHAEL MOORCOCK

horses. Licensed pearlies pause for a photo-opportunity in the dockside Secure Zones. Without all that cheap scenery, without our myths and magical skills, without our whorish good cheer and instincts for trade—any kind of trade—we probably wouldn't have a living city.

As it is, the real city I live in has per square inch more creative energy at work at any given moment than anywhere else on the planet. But you'd never know it from a stroll up the Strand. It's almost all in those lively little sidestreets the English-speaking tourists can't help feeling a bit nervous about and which the French adore.

If you use music for comfortable escape you'd probably find more satisfying and cheaper relief in a massage parlour than at the umpteenth revival of *The Sound of Music*. I'd tell that to any hesitant punter who's not too sure. Check out the phone boxes for the ladies, I'd say, or you can go to the half-price ticket-booth in Leicester Square and pick up a ticket that'll deliver real value—Ibsen or Shakespeare, Shaw or Greenbank. Certainly you can fork out three hundred sheets for a fifty-sheet ticket that in a justly ordered world wouldn't be worth two pee and have your ears salved and your cradle rocked for two hours. Don't worry, I'd tell them, I make no judgements. Some hard-working whore profits, whatever you decide. So who's the cynic?

I went on one of those tours when my friends Dave and Di from Bury came up for the Festival of London in 2001 and it's amazing the crap they tell people. They put sex, violence and money into every story. They know fuck-all. They soup everything up. It's *Sun*-reader history. Even the Beefeaters at the Tower. Poppinsland. All that old English duff.

It makes you glad to get back to Soho.

Not so long ago you would usually find me in the Princess Louise, Berwick Street, at lunch time, a few doors down from the Chinese chippy and just across from Mrs White's trim stall in Berwick Market. It's only a narrow door and is fairly easy to miss. It has one bottle-glass window onto the street. This is a public house which has not altered since the 1940s when it was very popular with Dylan Thomas, Mervyn Peake, Ruthven Todd, Henry Treece and a miscellaneous bunch of other Welsh adventurers who threatened for a while to take over English poetry from the Irish.

It's a shit pub, so dark and smoky you can hardly find your glass in front of your face, but the look of it keeps the tourists out. It's used by all the culture pros—from arty types with backpacks, who do specialised walking tours, to famous gallery owners and top museum management—and by the heavy metal bikers. We all get on a treat. We are mutually dependent in our continuing resistance to invasion or change, to the preservation of the best and most vital aspects of our culture. We leave them alone because they protect us from the tourists, who might recognise us and make us put on our masks in a hurry. They leave us alone because the police won't want to bother a bunch of well-connected middle-class wankers like us. It is a wonderful example of mutuality. In the back rooms, thanks to some freaky accoustics, you can talk easily above the music and hardly know it's there.

Over the years there have been some famous friendships and unions struck between the two groups. My own lady wife was known as Karla the She Goat in an earlier incarnation and had the most exquisite and elaborate tattoos I ever saw. She was a wonderful wife and would have made a perfect mother. She died on the A1, on the other side of Watford Gap. She had just found out she was pregnant and was making her last sentimental run. It did me in for marriage after that. And urban romance.

I first heard about London Bone in the Princess Lou when Claire Rood, that elegant old dyke from the Barbican, who'd tipped me off about my new tailor, pulled my ear to her mouth and asked me in words of solid gin and garlic to look out for some for her, darling. None of the usual faces seemed to know about it. A couple of top-level museum people knew a bit, but it was soon obvious they were hoping I'd fill them in on the details. I showed them a confident length of cuff. I told them to keep in touch.

I did my Friday walk, starting in the horrible pre-dawn chill of the Portobello Road where some youth tried to sell me a bit of scrimshawed reconstitute as 'the real old Bone.' I warmed myself in the showrooms of elegant Kensington and Chelsea dealers telling outrageous stories of deals, profits and crashes until they grew uncomfortable and wanted to talk about me and I got the message and left.

I wound up that evening in the urinal of The Dragoons in Meard Alley, swapping long-time-no-sees with my boyhood friend Bernie Michaud who begins immediately by telling me he's got a bit of business I might be interested in. And since it's Bernie Michaud telling me about it I listen. We settled down in a quiet corner of the pub. Bernie never deliberately spread a rumour in his life but he's always known how to make the best of one. This is kosher, he thinks. It has a bit of a glow. It smells like a winner. A long-distance runner. He is telling me out of friendship, but I'm not really interested. I'm trying to find out about London Bone.

"I'm not talking drugs, Ray, you know that. And it's not bent." Bernie's little pale face is serious. He takes a thoughtful sip of his whisky. "It is, admittedly, a commodity."

I wasn't interested. I hadn't dealt in goods for years. "Services only, Bernie," I said. "Remember. It's my rule. Who wants to get stuck paying rent on a warehouse full of yesterday's faves? I'm still trying to move those *Glenda Sings Michael Jackson* sides Pratface talked me into."

"What about investment?" he says. "This is the real business, Ray, believe me."

So I heard him out. It wouldn't be the first time Bernie had brought me back a nice profit on some deal I'd helped him bankroll and I was all right at the time. I'd just made the better part of a month's turnover on a package of theatreland's most profitable stinkers brokered for a party of filthy-rich New Muscovites who thought Chekhov was something you did with your lottery numbers.

As they absorbed the quintessence of Euro-ersatz, guaranteed to offer, as its high emotional moment, a long, relentless bowel movement, I would be converting their hard roubles back into beluga.

It's a turning world, the world of the international free market and everything's wonderful and cute and pretty and *magical* so long as you keep your place on the carousel. It's not good if it stops. And it's worse if you get thrown off altogether. Pray to Mammon that you never have to seek the help of an organization that calls you a 'client.' That puts you outside the fairground forever. No more rides. No more fun. No more life.

Bernie only did quality art, so I knew I could trust that side of his judgement, but what was it? A new batch of Raphaels turned up in a Willsden attic? Andy Warhol's lost landscapes found at the Pheasantry?

"There's American collectors frenzied for this stuff," murmurs Bernie through a haze of Sons of the Wind, Motorchair and Montecristo fumes. "And if it's decorated they go through the roof. All the big Swiss guys are looking for it. Freddy K in Cairo has a Saudi buyer who tops any price. Rose Sarkissian in Agadir represents three French collectors. It's never catalogued. It's all word of mouth. And it's already turning over millions. There's one inferior piece in New York and none at all in Paris. The pieces in Zurich are probably all fakes."

This made me feel that I was losing touch. I still didn't know what he was getting at.

"Listen," I say, "before we go any further, let's talk about this London Bone."

"You're a fly one, Ray," he says. "How did you suss it?"

"Tell me what you know," I say. "And then I'll fill you in."

We went out of the pub, bought some fish-and-chips at the Chinese and then walked up Berwick Street and round to his little club in D'Arblay Street where we sat down in his office and closed the door. The place stank of cat pee. He doted on his Persians. They were all out in the club at the moment, being petted by the patrons.

"First," he says, "I don't have to tell you, Ray, that this is strictly double-schtum and I will kill you if a syllable gets out."

"Naturally," I said.

"Have you ever seen any of this Bone?" he asked. He went to his cupboard and found some vinegar and salt. "Or better still, handled it?"

"No," I said. "Not unless it's fake scrimshaw."

"This stuff's got a depth to it you've never dreamed about. A lustre. You can tell it's the real thing as soon as you see it. Not just the shapes or the decoration, but the quality of it. It's like it's got a soul. You could come close, but you could never fake it. Like amber, for instance. That's why the big collectors are after it. It's authentic, it's newly discovered, and it's rare."

"What bone is it?"

"Mastodon. Some people still call it mammoth ivory, but I haven't seen any actual ivory. It could be dinosaur. I don't know. Anyway, this bone is *better* than ivory. It's in weird shapes, probably fragments off some really big animal."

"And where's it coming from?"

"The heavy clay of good old London," says Bernie. "A fortune at our feet, Ray. And my people know where to dig."

TWO

I had to be straight with Bernie. Until I saw a piece of the stuff in my own hand and got an idea about it for myself, I couldn't do anything. The only time in my life I'd gone for a gold brick I'd bought it out of respect for the genius running the scam. He deserved what I gave him. Which was a bit less than he was hoping for. Rather than be conned, I would throw the money away. I'm like that with everything.

I had my instincts, I told Bernie. I had to go with them. He understood completely and we parted on good terms.

If the famous Lloyd Webber meltdown of '03 had happened a few months earlier or later I would never have thought again about going into the Bone business, but I was done in by one of those sudden changes of public taste which made the George M. Cohan crash of '31 seem like a run of *The Mousetrap*.

Sentimental fascism went out the window. Liberal-humanist contemporary relevance, artistic aspiration, intellectual and moral substance and all that stuff was somehow in demand. It was *better* than the sixties. It was one of those splendid moments when the public pulls itself together and tries to grow up. Jones's *Rhyme of the Flying Bomb* song cycle made a glorious comeback. *American Angels* returned with even more punch. And Sondheim made an incredible comeback.

He became a quality brand-name. If it wasn't by Sondheim or based on a tune Sondheim used to hum in the shower, the punters didn't want to know. Overnight, the public's product loyalty had changed. And I must admit it had changed for the better. But my investments were in *Cats*, and *Dogs* (Lord Webber's last desperate attempt to squeeze from Thurber what he sucked from Eliot), *Duce!* and *Starlight Excess*, all of which were now taking a walk down *Sunset Boulevard*. I couldn't even get a regular price ticket for myself at *Sunday in the*

Park, *Assassins* or *Follies*. *Into The Woods* was solid for eighteen months ahead. I saw *Passion* from the wings and *Sweeney Todd* from the gods. *Five Guys Named Mo* crumbled to dust. *Phantom* closed. Its author claimed sabotage.

"Quality will out, Ray," says Bernie next time I see him at the Lou. "You've got to grant the public that. You just have to give it time."

"Fuck the public," I said, with some feeling. "They're just nostalgic for quality at the moment. Next year it'll be something else. Meanwhile I'm bloody ruined. You couldn't drum a couple of oncers on my entire stock. Even my ENO side-bets have died. Covent Garden's a disaster. The weather in Milan didn't help. That's where Cecilia Bartoli caught her cold. I was lucky to be offered half-price for the Rossinis without her. And I know what I'd do if I could get a varda at bloody Simon Rattle."

"So you won't be able to come in on the Bone deal?" said Bernie, returning to his own main point of interest.

"I said I was ruined," I told him, "not wiped out."

"Well, I got something to show you now, anyway," says Bernie.

We went back to his place.

He put it in my hand as if it were a nugget of plutonium, a knuckle of dark, golden Bone, split off from a larger piece, covered with tiny pictures.

"The engravings are always on that kind of Bone," he said. "There are other kinds that don't have drawings, maybe from a later date. It's the work of the first Londoners, I suppose, when it was still a swamp. About the time your Phoenician ancestors started getting into the upriver woad-trade. I don't know the significance, of course."

The Bone itself was hard to analyse because of the mixture of chemicals which has created it, and some of it had fused, suggesting prehistoric upheavals of some kind. The drawings were extremely primitive. Any bored person with a sharp object and minimum talent could have done them at any time in history. The larger, weirder-looking Bones, had no engravings.

Stick people pursued other stick people endlessly across the fragment. The work was unremarkable. The beauty really was in the tawny ivory colour of the Bone alone. It glowed with a wealth of shades and drew you hypnotically into its depths. I imagined the huge animal of which this fragment had once been an active part. I saw the bellowing trunk, the vast ears, the glinting tusks succumbing suddenly to whatever had engulfed her. I saw her body swaying, her tail lashing as she trumpeted her defiance of her inevitable death. And now men sought her remains as treasure. It was a very romantic image and of course it would become my most sincere sales pitch.

"That's six million dollars you're holding there," said Bernie. "Minimum."

Bernie had caught me at the right time and I had to admit I was convinced. Back in his office he sketched out the agreement. We would go in on a fifty-fifty basis, funding the guys who would do the actual digging, who knew where the Bone-fields were and who would tell us as soon as we showed serious interest. We would finance all the work, pay them an upfront earnest and then load by load in agreed increments. Bernie and I would split the net profit fifty-fifty. There were all kinds of clauses and provisions covering the various problems we foresaw, and then we had a deal.

The archeologists came round to my little place in Dolphin Square. They were a scruffy bunch of students from the University of Norbury who had discovered the Bone deposits on a run-of-the-mill field trip in a demolished Southwark housing estate and knew only that there might be a market for them. Recent cuts to their grants had made them desperate. Some lefty had come up with a law out of the Magna Carta, or whatever, saying public land couldn't be sold to private developers. Now there was a court case disputing the council's right to sell the estate to Livingstone International. This also put a stop to the planned rebuilding, so we had indefinite time to work.

The stoodies were grateful for our expertise, as well as our cash. I was happy enough with the situation. It was one I felt we could easily control. Middle-class burbnerds get greedy the same as anyone else, but they respond well to reason. I told them for a start-off that all the Bone had to come in to us. If any of it leaked onto the market by other means, we'd risk losing our prices and that would mean the scheme was over. Terminated, I said significantly. Since we had reputations as well as investments to protect there would also be recriminations. That's all I had to say. Since those V-serials kids think we're Krays and Mad Frankie Frasers just because we like to look smart and talk properly.

We were fairly sure we weren't doing anything obviously criminal. The stuff wasn't treasure trove. It had to be cleared before proper foundations could be poured. Quite evidently LI didn't think it was worth paying security staff to shuft the site. We didn't know if digging shafts and tunnels was even trespass, but we knew we had a few weeks before someone started asking about us and by then we hoped to have the whole bloody mastodon out of the deep clay and nicely earning for us. The selling would take the real skill and that was my job. It was going to have to be played sharper than South African diamonds.

After that neither Bernie nor I had anything to do with the dig. We rented a guarded lockup in Clapham and paid the kids every time they brought in a substantial load of Bone. It was incredible stuff. Bernie thought that chemical action, some of it relatively recent, had caused the phenomenon. "Like chalk, you know. You hardly find it anywhere. Just a few places in England, France, China and Texas."

The kids reported that there was more than one kind of animal down there, but that all the Bone had the same rich appearance. They had constructed a new tunnel, with a hidden entrance, so that even if the building site was blocked to them, they could still get at the Bone. It seemed to be a huge field, but most of the Bone was at roughly the same depth. Much of it had fused and had to be chipped out. They had found no end to it so far and they had tunneled through more than half an acre of the dense, dark clay.

Meanwhile I was in Amsterdam and Rio, Paris and Vienna and New York and Sydney. I was in Tokyo and Seoul and Hong Kong. I was in Riyadh, Cairo and Baghdad. I was in Kampala and New Benin, everywhere there were major punters. I racked up so many free airmiles in a couple of months that they were automatically jumping me to first class. But I achieved what I wanted. Nobody bought London Bone without checking with me. I was the acknowledged expert. The prime source, the best in the business. If you want Bone, said the art world, you want Gold.

The Serious Fraud Squad became interested in Bone for a while, but they had been assuming we were faking it and gave up when it was obviously not rubbish.

Neither Bernie nor I expected it to last any longer than it did. By the time we drew a line under our first phase of selling, we were turning over so much dough it was silly and the kids were getting tired and were worrying about exploring some of their wildest dreams. There was almost nothing left, they said. So we closed down the operation, moved our warehouses a couple of times and then let the Bone sit there to make us some money while everyone wondered why it had dried up.

And at that moment, inevitably, and late as ever, the newspapers caught on to the story. There was a brief late-night TV piece. A few supplements talked about it in their arts pages. This led to some news stories and eventually it went to the tabloids and became anything you liked from the remains of Martians to a new kind of nuclear waste. Anyone who saw the real stuff was convinced but everyone had a theory about it. The real exclusive market was finished. We kept schtum. We were gearing up for the second phase. We got as far away from our stash as possible.

Of course a few faces tracked me down, but I denied any knowledge of the Bone. I was a middle-man, I said. I just had good contacts. Half-a-dozen people claimed to know where the Bone came from. Of course they talked to the papers. I sat back in satisfied security, watching the mud swirl over our tracks. Another couple of months and we'd be even safer than the house I'd bought in Hampstead overlooking the Heath. It had a rather forlorn garden the size of Kilburn which needed a lot of nurturing. That suited me. I was ready to retire to the country and a big indoor swimming pool.

By the time a close version of the true story came out, from one of the stoodies who'd lost all his share in a lottery syndicate, it was just one of many. It sounded too dull. I told newspaper reporters that while I would love to have been involved in such a lucrative scheme, my money came from theatre tickets. Meanwhile, Bernie and I thought of our warehouse and said nothing.

Now the stuff was getting into the culture. It was chic. *Puncher* used it in their ads. It was called Mammoth Bone by the media. There was a common story about how a herd had wandered into the swampy river and drowned in the mud. Lots of pictures dusted off from the Natural History Museum. Experts explained the colour, the depths, the markings, the beauty. Models sported a Bone motif.

Our second phase was to put a fair number of inferior fragments on the market and see how the public responded. That would help us find our popular price—the most a customer would pay. We were looking for a few good millionaires.

Frankly, as I told my partner, I was more than ready to get rid of the lot. But Bernie counselled me to patience. We had a plan and it made sense to stick to it.

The trade continued to run well for a while. As the sole source of the stuff, we could pretty much control everything. Then one Sunday lunchtime I met Bernie at The Six Jolly Dragoons in Meard Alley, Soho. He had something to show me, he said. He didn't even glance around. He put it on the bar in plain daylight. A small piece of Bone with the remains of decorations still on it.

"What about it?" I said.

"It's not ours," he said.

My first thought was that the stoodies had opened up the field again. That they had lied to us when they said it had run out.

"No," said Bernie, "it's not even the same colour. It's the same stuff—but different shades. Gerry Goldstein lent it to me."

"Where did he get it?"

"He was offered it," he said.

We didn't bother to speculate where it had come from. But we did have rather a lot of our Bone to shift quickly. Against my will, I made another world tour and sold mostly to other dealers this time. It was a standard second-wave operation but run rather faster than was wise. We definitely missed the crest.

However, before deliveries were in and cheques were cashed, Jack Merrywidow, the fighting MP for Brookgate and E. Holborn, gets up in the House of Commons during telly-time one afternoon and asks if Prime Minister Bland or any of his dope-dazed cabinet understand that human remains, taken from the hallowed burial grounds of London, are being sold by the piece in the international marketplace? Mr Bland makes a plummy joke enjoyed at Mr Merrywidow's expense and sits down. But Jack won't give up. Next week he's back on telly for the *Struggle of Parliament* interview. Jack's had the Bone examined by experts. It's human. Undoubtedly human. The strange shapes are caused by limbs melting together in soil heavy with lime. Chemical reactions, he says. We have—he raises his eyes to the camera—been mining mass graves.

A shock to all those who still long for the years of common decency. Someone, says Jack, is mining more than our heritage. Hasn't free market capitalism got a little bit out of touch when we start selling the arms, legs and skulls of our forebears? The torsos and shoulder-blades of our honourable dead? What did we use to call people who did that? When was the government going to stop this trade in corpses?

It's denied.

It's proved.

It looks like trade is about to slump.

I think of framing the cheques as a reminder of the vagaries of fate and give up any idea of popping the question to my old muse Little Trudi, who is back on the market, having been dumped by her corporate suit in a fit, he's told her, of self-disgust after seeing *The Tolstoy Investment* with Eddie Izzard. Bernie, I tell my partner, the Bone business is down the drain. We might as well bin the stuff we've stockpiled.

Then two days later the TV news reports a vast public interest in London Bone. Some lordly old queen with four names comes on the evening news to say how by owning a piece of Bone, you own London's true history. You become a curator of some ancient ancestor. He's clearly got a vested interest in the stuff. It's the hottest tourist item since Jack the Ripper razors and OJ gloves. More people want to buy it than ever.

The only trouble is, I don't deal in dead people. It is, in fact, where I have always drawn the line. Even Pratface Charlie wouldn't sell his great-great-grandmother's elbow to some overweight Jap in a deerstalker and a kilt. I'm faced with a genuine moral dilemma.

I make a decision. I make a promise to myself. I can't go back on that. I go down to the Italian chippy in Fortess Road, stoke up on nourishing ritual grease (cod, roe, chips and mushy peas, bread and butter and tea, syrup pudding), then heave my out-of-shape, but mentally prepared, body up onto Parliament Hill to roll myself a big wacky-baccy fag and let my subconscious think the problem through.

When I emerge from my reverie, I have looked out over the whole misty London panorama and considered the city's complex history. I have thought about the number of dead buried there since, say, the time of Boadicea, and what they mean to the soil we build on, the food we still grow here and the air we breathe. We are recycling our ancestors all the time, one way or another. We are sucking them in and shitting them out. We're eating them. We're drinking them. We're coughing them up. The dead don't rest. Bits of them are permanently at work. So what am I doing wrong?

This thought is comforting until my moral sense, sharpening itself up after a long rest, kicks in with—but what's different here is you're flogging the stuff to people who take it home with them. Back to Wisconsin and California and Peking. You take it out of circulation. You're dissipating the deep fabric of the city. You're unravelling something. Like, the real infrastructure, the spiritual and physical bones of an ancient settlement...

On Kite Hill I suddenly realise that those bones are in some way the deep lifestuff of London.

It grows dark over the towers and roofs of the metropolis. I sit on my bench and roll myself up a further joint. I watch the silver rising from the river, the deep golden glow of the distant lights, the plush of the foliage, and as I watch it seems to shred before my eyes, like a rotten curtain. Even the traffic noise grows fainter. Is the city sick? Is she expiring? Somehow it seems there's a little less breath in the old girl. I blame myself. And Bernie. And those kids.

There and then, on the spot, I renounce all further interest in the Bone trade. If nobody else will take the relics back, then I will.

There's no resolve purer than that which you draw from a really good reefer.

THREE

So now there isn't a tourist in any London market or antique arcade who isn't searching out Bone. They know it isn't cheap. They know they have to pay. And pay they do. Through the nose. And half of what they buy is crap or fakes. This is a question of status, not authenticity. As long as we say it's good, they can say it's good. We give it a provenance, a story, something to colour the tale to the folks back home. We're honest dealers. We sell only the authentic stuff. Still they get conned. But still they look. Still they buy.

Jealous Mancunians and Brummies long for a history old enough to provide them with Bone. A few of the early settlements, like Chester and York, start turning up something like it, but it's not the same. Jim Morrison's remains disappear from Pére La Chaise. They might be someone else's bones, anyway. Rumour is they were KFC bones. The revolutionary death-pits fail to deliver the goods. The French are furious. They accuse the British of gross materialism and poor taste. Oscar Wilde disappears. George Eliot. Winston Churchill. You name them. For a few months there is a grotesque trade in the remains of the famous. But the fashion has no intrinsic substance and fizzles out. Anyone could have seen it wouldn't run.

Bone has the image, because Bone really is beautiful.

Too many people are yearning for that Bone. The real stuff. It genuinely hurts me to disappoint them. Circumstances alter cases. Against my better judgement I continue in the business. I bend my principles, just for the duration. We have as much turnover as we had selling to the Swiss gnomes. It's the latest item on the *beento* list. "You *have* to bring me back some London Bone, Ethel, or I'll never forgive you!" It starts to appear in the American luxury catalogs.

But by now there are ratsniffers everywhere—from Trade and Industry, from the National Trust, from the Heritage Corp, from half-a-dozen South London councils, from the Special Branch, from the CID, the Inland Revenue and both the Funny and the Serious Fraud Squads.

Any busybody who ever wanted to put his head under someone else's bed is having a wonderful time. Having failed dramatically with the STOP THIS DISGUSTING TRADE approach, the tabloids switch to offering bits of Bone as prizes in circulation boosters. I sell a newspaper consortium a Tesco's plastic bagfull for two-and-a-half mill via a go-between. Bernie and I are getting almost frighteningly rich. I open some bank accounts offshore and I become an important anonymous shareholder in the Queen Elizabeth Hall when it's privatized.

It doesn't take long for the experts to come up with an analysis. Most of the Bone has been down there since the seventeenth century and earlier. They are the sites of the old plague pits where legend had it still-living bodies were thrown in with the dead. For a while it must have seemed like Auschwitz-on-Thames. The chemical action of lime, partial burning, London clay and decaying flesh, together with the broadening spread of the London water-table, thanks to various engineering works over the last century, letting untreated sewage into the mix, had created our unique London Bone. As for the decorations, that, it was opined, was the work of the pit guards, working on earlier bones found on the same site.

"Blood, shit and bone," says Bernie. "It's what makes the world go round. That and money, of course."

"And love," I add. I'm doing all right these days. It's true what they say about a Roller. Little Trudi has enthusiastically rediscovered my attractions. She has her eye on a ring. I raise my glass. "And love, Bernie."

"Fuck that," says Bernie. "Not in my experience." He's buying Paul McCartney's old place in Wamering and having it converted for Persians. He has, it is true, also bought his wife her dream house. She doesn't seem to mind it's on the island of Las Cascadas about six miles off the coast of Morocco. She's at last agreed to divorce him. Apart from his mother, she's the only woman he ever had anything to do with and he isn't, he says, planning to try another. The only females he wants in his house in future come with a pedigree a mile long, have all their shots and can be bought at Harrods.

FOUR

I expect you heard what happened. The private Bonefields, which contractors were discovering all over South and West London, actually contained public bones. They were part of our national inheritance. They had living relatives. And stones, some of them. So it became a political and a moral issue. The Church got involved. The airwaves were crowded with concerned clergy. There was the problem of the self-named bone-miners. Kids, inspired by our leaders' rhetoric, and aspiring to imitate those great captains of free enterprise they had been taught to admire, were turning over ordinary graveyards, which they'd already stripped of their saleable masonry, and digging up somewhat fresher stiffs than was seemly.

A bit too fresh. It was pointless. The Bone took centuries to get seasoned and so far nobody had been able to fake the process. A few of the older graveyards had small deposits of Bone in them. Brompton Cemetery had a surprising amount, for instance, and so did Highgate. This attracted prospectors. They used shovels mainly, but sometimes low explosives. The area around Karl Marx's monument looked like they'd refought the Russian Civil War over it. The barbed wire put in after the event hadn't helped. And as usual the public paid to clean up after private enterprise. Nobody in their right mind got buried any more. Cremation became very popular. The borough councils and their financial managers were happy because more valuable real estate wasn't being occupied by a non-consumer.

It didn't matter how many security guards were posted or, by one extreme Authority, land-mines: the teenies left no grave unturned. Bone was still a profitable item, even though the market had settled down since we started. They dug up Bernie's mother. They dug up my cousin Leonard. There wasn't a Londoner who didn't have some intimate unexpectedly back above ground. Every night you saw it on telly.

It had caught the public imagination. The media had never made much of the desecrated graveyards, the chiselled-off angels' heads and the uprooted headstones on sale in King's Road and the Boulevard St. Michel since the 1970s. These had been the targets of first-generation grave-robbers. Then there had seemed nothing left to steal. Even they had balked at doing the corpses. Besides, there wasn't a market. This second generation was making up for lost time, turning over the soil faster than an earthworm on E.

The news shots became clichés. The heaped earth, the headstone, the smashed coffin, the hint of the contents, the leader of the Opposition coming on to say how all this has happened since his mirror image got elected. The councils argued that they should be given the authority to deal with the problem. They owned the graveyards. And also, they reasoned, the Bonefields. The profits from those fields should rightly go into the public purse. They could help pay for the Health Service. "Let the dead," went their favourite slogan, "pay for the living for a change."

What the local politicians actually meant was that they hoped to claim the land in the name of the public and then make the usual profits privatising it. There was a principle at stake. They had to ensure their friends and not outsiders got the benefit.

The High Court eventually gave the judgement to the public, which really meant turning it over to some of the most rapacious borough councils in our history. In the 1980s, that Charlie Peace of elected bodies, the Westminster City Council, had tried to sell their old graveyards to new developers. This current judgement allowed all councils at last to maximise their assets from what was, after all, dead land, completely unable to pay for itself, and therefore a natural target for privatisation. The feeding frenzy began. It was the closest thing to mass cannibalism I've ever seen.

We had opened a fronter in Old Sweden Street and had a couple of halfway presentable slags from Bernie's club taking the calls and answering enquiries. We were straight up about it. We called it *The City Bone Exchange*. The bloke who decorated it and did the sign specialised in giving offices that long-established look. He'd created most of those old-fashioned West End Hotels you'd never heard of until 1999. "If it's got a Scottish name," he used to say, "it's one of mine. Americans love the skirl of the pipes, but they trust a bit of brass and varnish best."

Our place was almost all brass and varnish. And it worked a treat. The Ritz and the Savoy sent us their best potential buyers. Incredibly exclusive private hotels gave us taxi-loads of bland-faced American boy-men, reeking of health and beauty products and bellowing their credentials to the wind, rich matrons eager for anyone's approval, massive Germans with aggressive cackles, stern orientals glaring at us, daring us to cheat them. They bought. And they bought. And they bought.

The snoopers kept on snooping but there wasn't really much to find out. Livingstone International took an aggressive interest in us for a while, but what could they do? We weren't up to anything illegal just selling the stuff and nobody could identify what if anything had been nicked anyway. I still had my misgivings. They weren't anything but superstitions, really. It did seem sometimes that for every layer of false antiquity, for every act of disneyfication, an inch or two of our real foundations crumbled. You knew what happened when you did that to a house. Sooner or later you got trouble. Sooner or later you had no house.

We had more than our share of private detectives for a while. They always pretended to be customers and they always looked wrong, even to our girls. Livingstone International had definitely made a connection. I think they'd found our mine and guessed what a windfall they'd lost. They didn't seem at one with themselves over the matter. They even made veiled threats. There was some swagger came in to talk about violence but they were spotties who'd got all their language off old nineties TV shows. So we sweated it out and the girls took most of the heat. Those girls really didn't know anything. They were magnificently ignorant. They had tellies with chips which switch channels as soon as they detect a news or information programme.

I've always had a rule. If you're caught by the same wave twice, get out of the water.

While I didn't blame myself for not anticipating the Great Andrew Lloyd Webber Slump, I think I should have guessed what would happen next. The tolerance of the public for bullshit had become decidedly and aggressively negative. It was like the Bone had set new standards of public aspiration as well as beauty. My dad used to say that about the Blitz. Classical music enjoyed a huge success during the Second World War. Everybody grew up at once. The Bone had made it happen again. It was a bit frightening to those of us who had always relied on a nice, passive, gullible, greedy punter for an income.

The bitter fights which had developed over graveyard and Bonefield rights and boundaries, the eagerness with which some borough councils exploited their new resource, the unseemly trade in what was, after all, human remains, the corporate involvement, the incredible profits, the hypocrisies and politics around the Bone brought us the outspoken disgust of Europe. We were used to that. In fact, we tended to cultivate it. But that wasn't the problem.

The problem was that our *own* public had had enough.

When the elections came round, the voters systematically booted out anyone who had supported the Bone trade. It was like the sudden rise of the anti-slavery vote in Lincoln's America. They demanded an end to the commerce in London Bone. They got the Boneshops closed down. They got work on the Bonefields stopped. They got their graveyards and monuments protected and cleaned up. They got a city which started cultivating peace and security as if it was a cash crop. Which maybe it was. But it hurt me.

It was the end of my easy money, of course. I'll admit I was glad it was stopping. It felt like they were slowing entropy, restoring the past. The quality of life improved. I began to think about letting a few rooms for company.

The mood of the country swung so far into disapproval of the Bone trade that I almost began to fear for my life. Road and anti-abortion activists switched their attention to Bone merchants. Hampstead was full of screaming lefties convinced they owned the moral highground just because they'd paid off their enormous mortgages. Trudi, after three months, applied for a divorce, arguing that she had not known my business when she married me. She said she was disgusted. She said I'd been living on blood-money. The courts awarded her more than half of what I'd made, but it didn't matter any more. My investments were such that I couldn't stop earning. Economically, I was a small oil-producing nation. I had my own international dialling code. It was horrible in a way. Unless I tried very hard, it looked like I could never be ruined again. There was no justice.

I met Bernie in *The King Lyar* in Old Sweden Street, a few doors down from our burned-out office. I told him what I planned to do and he shrugged.

"We both knew it was dodgy," he told me. "It was dodgy all along, even when we thought it was mastodons. What it feels like to me, Ray, is—it feels like a sort of a massive transformation of the *zeitgeist*—you know, like Virginia Woolf said about the day human nature changed—something happens slowly and you're not aware of it. Everything seems normal. Then you wake up one morning and—bingo!—it's Nazi Germany or Bolshevik Russia or Thatcherite England or the Golden Age—and all the rules have changed."

"Maybe it was the Bone that did it," I said. "Maybe it was a symbol everyone needed to rally round. You know. A focus."

"Maybe," he said. "Let me know when you're doing it. I'll give you a hand."

About a week later we got the van backed up to the warehouse loading bay. It was three o'clock in the morning and I was chilled to the marrow. Working in silence we transferred every scrap of Bone to the van. Then we drove back to Hampstead through a freezing rain.

I don't know why we did it the way we did it. There would have been easier solutions, I suppose. But behind the high walls of my big back garden, under the old trees and etiolated rhododendrons, we dug a pit and filled it with the glowing remains of the ancient dead.

The stuff was almost phosphorescent as we chucked the big lumps of clay back on to it. It glowed a rich amber and that faint, rosemary smell came off it. I can still smell it when I go in there to this day. My soft fruit is out of this world. The whole garden's doing wonderfully now.

In fact London's doing wonderfully. We seem to be back on form. There's still a bit of a Bone trade, of course, but it's marginal.

Every so often I'm tempted to take a spade and turn over the earth again, to look at the fortune I'm hiding there. To look at the beauty of it. The strange amber glow never fades and sometimes I think the decoration on the Bone is an important message I should perhaps try to decipher.

I'm still a very rich man. Not justly so, but there it is. And, of course, I'm about as popular with the public as Percy the Paedophile. Gold the Bone King? I might as well be Gold the Grave Robber. I don't go down to Soho much. When I do make it to a show or something I try to disguise myself a bit. I don't see anything of Bernie any more and I heard two of the stoodies topped themselves.

I do my best to make amends. I'm circulating my profits as fast as I can. Talent's flooding into London from everywhere, making a powerful mix. They say they haven't known a buzz like it since 1967. I'm a reliable investor in great new shows. Every year I back the Iggy Pop Awards, the most prestigious in the business. But not everybody will take my money. I am regularly reviled. That's why some organisations receive anonymous donations. They would refuse them if they knew they were from me.

I've had the extremes of good and bad luck riding this particular switch in the zeitgeist and the only time I'm happy is when I wake up in the morning and I've forgotten who I am. It seems I share a common disgust for myself.

A few dubious customers, however, think I owe them something.

Another bloke, who used to be very rich before he made some frenetic investments after his career went down the drain, called me the other day. He knew of my interest in the theatre, that I had invested in several West End hits. He thought I'd be interested in his idea. He wanted to revive his first success, *Rebecca's Incredibly Far*

Out Well or something, which he described as a powerful religious rock opera guaranteed to capture the new nostalgia market. The times, he told me, they were a-changin'. His show, he continued, was full of raw old-fashioned R&B energy. Just the sort of authentic sound to attract the new no-nonsense youngsters. Wasn't it cool that Madonna wanted to do the title role? And Bob Geldof would play the Spirit of the Well. *Rock and roll, man! It's all in the staging, man! Remember the boat in Phantom? I can make it look better than real. On stage, man, that well is W.E.T. WET! Rock and roll!* I could see that little wizened fist punching the air in a parody of the vitality he craved and whose source had always eluded him.

I had to tell him it was a non-starter. I'd turned over a new leaf, I said. I was taking my ethics seriously.

These days I only deal in living talent.

THIRTEEN VIEWS OF A CARDBOARD CITY

BY WILLIAM GIBSON

ONE
DEN-EN

Low angle, deep perspective, establishing Tokyo subway station interior.

Shot with available light, long exposure; a spectral pedestrian moves away from us, into background. Two others visible as blurs of motion.

Overhead fluorescents behind narrow rectangular fixtures. Ceiling tiled with meter-square segments (acoustic baffles?). Round fixtures are ventilators, smoke-detectors, speakers? Massive square columns recede. Side of a stairwell or escalator. Mosaic tile floor in simple large-scale pattern: circular white areas in square tiles, black infill of round tiles. The floor is spotless: no litter at all. Not a cigarette butt, not a gum-wrapper.

A long train of cardboard cartons, sides painted with murals, recedes into the perspective of columns and scrubbed tile: first impression is of a children's art project, something choreographed by an aggressively creative preschool teacher. But not all of the corrugated cartons have been painted; many, particularly those farthest away, are bare brown paper. The one nearest the camera, unaltered, bright yellow, bears the Microsoft logo.

The murals appear to have been executed in poster paints, and are difficult to interpret here.

There are two crisp-looking paper shopping-bags on the tile floor: one near the murals, the other almost in the path of the ghost pedestrian. These strike a note of anomaly, of possible threat: London Transport warnings, Sarin cultists... Why are they there? What do they contain?

The one nearest the murals bears the logo "DEN-EN."

Deeper in the image are other cartons. Relative scale makes it easier to see that these are composites, stitched together from smaller boxes. Closer study makes the method of fastening clear: two sheets are punctured twice with narrow horizontal slits, flat poly-twine analog (white or pink) is threaded through both sheets, a knot is tied, the ends trimmed neatly. In fact, all of the structures appear to have been assembled this way.

Deepest of all, stairs. Passengers descending.

TWO
BLUE OCTOPUS

Shallow perspective, eye-level, as though we were meant to view an anamorphic painting.

This structure appears to have been braced with a pale blue, enameled, possibly spring-loaded tube with a white, non-slip plastic foot. It might be the rod for a shower-curtain, but here it is employed vertically. Flattened cartons are neatly lashed to this with poly-tie.

The murals. Very faintly, on the end of the structure, nearest the camera, against a black background, the head of the Buddha floats above something amorphous and unreadable. Above the Buddha are fastened what appear to be two packaging-units for Pooh Bear dolls. These may serve a storage function. The mural on the face of the structure is dark, intricate, and executed (acrylic paints?) with considerable technique. Body parts, a sense of claustrophobic, potentially erotic proximity. A female nude, head lost where the cardboard ends, clutches a blue octopus whose tentacles drape across the forehead of a male who seems to squat doglike at her feet. Another nude lies on her back, knees upraised, her sex shadowed in perspective. The head of a man with staring eyes and pinprick pupils hovers above her ankles; he appears to be smoking but has no cigarette.

A third nude emerges, closest to the camera: a woman whose features suggest either China or the Mexico of Diego Rivera.

A section of the station's floor, the round black tiles, is partially covered with a scrap of grayish-blue synthetic pile carpeting.

Pinned eyes.

THREE
FRONTIER INTERNATIONAL

Shot straight back into what may be a wide alcove. Regular curves of pale square tiles.

Four structures visible.

The largest, very precisely constructed, very hard-edged, is decorated with an eerie pointillist profile against a solid black background: it seems to be a very old man, his chin, lipless mouth and drooping nose outlined in blood red. In front of this is positioned a black hard-sided overnighter suitcase.

Abutting this structure stands another, smaller, very gaily painted: against a red background with a cheerful yellow bird and yellow concentric circles, a sort of Cubist ET winks out at the camera. The head of a large nail or pin, rendered in a far more sophisticated style, penetrates the thing's forehead above the open eye.

A life-sized human hand, entirely out of scale with the huge head, is reaching for the eye.

Nearby sits an even smaller structure, this one decorated with abstract squares of color recalling Klee or Mondrian. Beside it is an orange plastic crate of the kind used to transport sake bottles. An upright beer can. A pair of plastic sandals, tidily arranged.

Another, bigger structure behind this one. Something painted large-scale in beige and blue (sky?) but this is obscured by the Mondrian. A working door, hinged with poly-tie, remains unpainted: the carton employed for the door is printed with the words "FRONTIER INTERNATIONAL."

Individual styles of workmanship start to become apparent.

Deeper in the image, beyond what appears to be a stack of neatly-folded blankets, is located the blue enamel upright, braced against the ceiling tile. Another like it, to its right, supports a paper kite with the printed face of a samurai.

FOUR
AFTER PICASSO

Shallow perspective of what appears to be a single, very narrow shelter approximately nine meters in length. Suggests the literally marginal nature of these constructions: someone has appropriated less than a meter at the side of a corridor, and built along it, tunneling like a cardboard seaworm.

The murals lend the look of a children's cardboard theater.

Punch in the underground.

Like so many of the anonymous paintings to be found in thrift shops everywhere, these murals are somehow vaguely after Picasso. Echo of *Guernica* in these tormented animal forms. Human features rendered flounder-style: more Oxfam Cubism.

Square black cushion with black tassels at its corners, top an uncharacteristically peaked section of cardboard roof. Elegant.

The wall behind the shelter is a partition of transparent lucite, suggesting the possibility of a bizarre ant-farm existence.

FIVE
YELLOW SPERM

We are in an impossibly narrow "alley" between shelters, perhaps a communal storage area. Cardboard shelving, folded blankets.

A primitive portrait of a black kitten, isolated on a solid green ground, recalls the hypnotic stare of figures in New England folk art.

Also visible: the white plastic cowl of an electric fan, yellow plastic sake crate, pale blue plastic bucket, section of blue plastic duck-board, green plastic dustpan suspended by string, child's pail in dark blue plastic. Styrofoam takeaway containers with blue and scarlet paint suggest more murals in progress.

Most striking here is the wall of a matte-black shelter decorated with a mural of what appear to be large yellow inner-tubes with regularly spaced oval "windows" around their perimeters; through each window is glimpsed a single large yellow sperm arrested in mid-wriggle against a nebulous black-and-yellow background.

SIX
GOMI GUITAR

Extreme close, perhaps at entrance to a shelter.

An elaborately designed pair of black-and-purple Nike trainers, worn but clean. Behind them a pair of simpler white Reeboks (a woman's?).

A battered acoustic guitar strung with nylon. Beside it, a strange narrow case made of blue denim, trimmed with red imitation leather; possibly a golf bag intended to carry a single club to a driving range?

A self-inking German rubber stamp.

Neatly folded newspaper with Japanese baseball stars.

A battered pump-thermos with floral design.

SEVEN
108

A space like the upper berths on the Norfolk & Western sleeping cars my mother and I took when I was a child. Form following function.

The structure is wide enough to accommodate a single traditional Japanese pallet. A small black kitten sits at its foot (the subject of the staring portrait?). Startled by the flash, it is tethered with a red leash. A second, larger tabby peers over a shopping bag made of tartan paper. The larger cat is also tethered, with a length of thin white poly rope.

Part of a floral area-rug visible at foot of bed.

This space is deeply traditional, utterly culture-specific.

Brown cardboard walls, cardboard mailing tubes used as structural uprights, the neat poly-tie lashings.

On right wall:

GIC

MODEL NO: VS-30

Q'TY : 1 SET

COLOR: BLACK

C/T NO: 108

MADE IN KOREA

At the rear, near what may be assumed to be the head of the bed, are suspended two white-coated metal shelves or racks. These contain extra bedding, a spare cat-leash, a three-pack of some pressurized product (butane for a cooker?), towels.

On the right wall are hung two pieces of soft luggage, one in dark green imitation leather, the other in black leather, and a three-quarter-length black leather car coat.

On the left wall, a white towel, a pair of bluejeans, and two framed pictures (content not visible from this angle).

A section of transparent plastic has been mounted in the ceiling to serve as a skylight.

EIGHT
HAPPY HOUR

Wall with mailing-tube uprights.

A large handbill with Japanese stripper: "LIVE NUDE," "TOPLESS BOTTOMLESS," "HAPPY HOUR." Menu-chart from a hamburger franchise illustrating sixteen choices.

Beneath these, along the wall, are arranged two jars containing white plastic spoons, a tin canister containing chopsticks, eight stacked blue plastic large takeaway cups, fourteen stacked white paper takeway cups (all apparently unused, and inverted to protect against dust), neatly folded towels and bedding, aluminum cookware, a large steel kettle, a pink plastic dishpan, a large wooden chopping-board.

Blanket with floral motif spread as carpet.

NINE
SANDY

A different view of the previous interior, revealing a storage loft very tidily constructed of mailing-tubes and flattened cartons.

The similarities with traditional Japanese post-and-beam construction is even more striking, here. This loft-space is directly above the stacked cookware in the preceding image. Toward its left side is a jumble of objects, some unidentifiable: heavy rope, a child's plaid suitcase, a black plastic bowl, a softball bat. To the right are arranged a soft, stuffed baby doll, a plush stuffed dog, a teddy bear wearing overalls that say "SANDY," what seems to be a plush stuffed killer whale (shark?) with white felt teeth. The whale or shark still has the manufacturer's cardboard label attached, just as it came from the factory.

In the foreground, on the lower level, is a stack of glossy magazines, a tin box that might once have held candy or some other confection, and an open case that probably once contained a pair of sunglasses.

TEN
BOY'S BAR KYOKA

A very simple shot, camera directed toward floor, documenting another food-preparation area.

A square section of the round tiles is revealed at the bottom of the photograph. The rest of the floor is covered by layers of newspaper beneath a sheet of brown cardboard. A narrow border of exposed newsprint advertises "Boy's Bar KYOKA."

A blue thermos with a black carrying-strap. A greasy-looking paper cup covered with crumpled aluminum foil. A red soap-dish with a bar of white soap. A cooking-pot with an archaic-looking wooden lid. The pot's handle is wrapped in a white terry face cloth, secured with two rubber bands. Another pot, this one with a device for attaching a missing wooden handle, contains a steel ladle and a wooden spatula. A nested collection of plastic mixing bowls and colanders.

A large jug of bottled water, snow-capped peaks on its blue and white label.

A white plastic cutting-board, discolored with use. A white plastic (paper?) bag with "ASANO" above a cartoon baker proudly displaying some sort of loaf.

ELEVEN
J.O.

The shelters have actually enclosed a row of pay telephones!

Dial 110 for police.

Dial 119 for fire or ambulance.

Two telephones are visible: they are that singularly bilious shade of green the Japanese reserve for pay phones. They have slots for phone-cards, small liquid crystal displays, round steel keys. They are mounted on individual stainless-steel writing-ledges, each supported by a stout, mirror-finished steel post. Beneath each ledge is an enclosed shelf or hutch, made of black, perforated steel sheeting. Provided as a resting place for a user's parcels.

The hutches now serve as food-prep storage: four ceramic soup bowls of a common pattern, three more with a rather more intricate glaze, four white plastic bowls and several colored ones. A plastic scrubbing-pad, used.

On the floor below, on newspaper, are an aluminum teapot and what may be a package of instant coffee sachets. Three liter bottles of cooking oils.

On the steel ledge of the left-hand phone is a tin that once contained J.O. Special Blend ready-to-drink coffee.

TWELVE
NIPPON SERIES

An office.

A gap has been left in the corrugated wall, perhaps deliberately, to expose a detailed but highly stylized map of Tokyo set into the station's wall. The wall of this shelter and the wall of the station have become confused. Poly-tie binds the cardboard house directly into the fabric of the station, into the Prefecture itself.

This is quite clearly an office.

On the wall around the official, integral subway map, fastened to granite composite and brown cardboard with bits of masking tape: a postcard with a cartoon of orange-waistcoated figures escorting a child through a pedestrian crossing, a restaurant receipt (?), a newspaper clipping, a small plastic clipboard with what seem to be receipts, possibly from an ATM, a souvenir program from the 1995 Nippon Series (baseball), and two color photos of a black-and-white cat. In one photo, the cat seems to be here, among the shelters.

Tucked behind a sheet of cardboard are four pens and three pairs of scissors. A small pocket flashlight is suspended by a lanyard of white poly-tie.

To the right, at right angles to the wall above, a cardboard shelf is cantilevered with poly-tie. It supports a box of washing detergent, a book, a dayglo orange Casio G-Shock wristwatch, a white terry face cloth, a red plastic AM/FM cassette-player, and three disposable plastic cigarette-lighters.

Below, propped against the wall, is something that suggests the bottom of an inexpensive electronic typewriter of the sort manufactured by Brother.

A box of Chinese candy, a cat-brush, a flea-collar.

THIRTEEN
TV SOUND

Close-up of the contents of the shelf.

The red stereo AM-FM cassette-player, its chrome antenna extended at an acute angle for better reception. It is TV Sound brand, model LX-43. Its broken handle, mended with black electrical tape, is lashed into the structure with white poly-tie. Beside the three lighters, which are tucked partially beneath the player, in a row, are an unopened moist towelette and a red fine-point felt pen. To the left of the player is a square red plastic alarm clock, the white face cloth, and the Casio G-Shock. The Casio is grimy, one of the only objects in this sequence that actually appears to be dirty. The book, atop the box of laundry detergent, is hardbound, its glossy dustjacket bearing the photograph of a suited and tied Japanese executive. It looks expensive. Inspirational? Autobiographical?

To the right of the LX-43: a rigid cardboard pack of Lucky Strike non-filters and a Pokka coffee tin with the top neatly removed (to serve as an ashtray?).

On the cardboard bulkhead above these things are taped up two sentimental postcards of paintings of kittens playing. "Cat collection" in a cursive font.

Below these are glued (not taped) three black-and-white photographs.

#1: A balding figure in jeans and a short-sleeved t-shirt squats before an earlier, unpainted version of this structure.

One of the cartons seems to be screened with the word "PLAST—". He is eating noodles from a pot, using chopsticks.

#2: The "alley" between the shelters. The balding man looks up at the camera. Somehow he doesn't look Japanese at all. He sits cross-legged among half-a-dozen others. They look Japanese. All are engrossed in something, perhaps the creation of murals.

#3: He squats before his shelter, wearing molded plastic sandals. His hands grip his knees. Now he looks entirely Japanese, his face a formal mask of suffering.

Curve of square tiles.

How long has he lived here?

With his cats, his guitar, his neatly folded blankets?

Dolly back.

Hold on the cassette player

Behind it, almost concealed, is a Filofax.

Names.

Numbers.

Held as though they might be a map, a map back out of the underground.

THE AUTHORS

PAT CADIGAN

is the author of *Mindplayers*, *Synners* and *Fools*, the last two of which won the Arthur C. Clarke Award (in 1992 and 1995) making her the only serial Clarke Award winner so far. After almost a quarter of a century in Kansas, in 1996 she packed up and moved to Harringay in North London, where she and her son Bobzilla, former scourge of the Mid-West and now Harringay's Hacker Hurricane, live with her most recent husband, the Original Chris Fowler. Even so, Cadigan insists that getting married is not a habit and she can quit any time she wants. 1998 will see publication of her new novel (Tor, USA; HarperCollins, UK), by which time everyone involved should have agreed on what the title should be. Cadigan finds the gritty urban ambiance of Harringay stimulating and evocative, and plans to write many more novels in between learning to count in Kurdish from the local store owners, dodging the traffic on Green Lane and visiting the crusties in the park.

ERIC BROWN

was born in 1960 in Haworth, West Yorkshire. He began writing at fifteen while living in Mordialloc, Australia, and completed twenty novels and countless short stories before making his first sale to *Interzone* in 1986. Since then he has sold two novels, *Meridian Days* (1992) and *Engineman* (1994), and two collections, *The Time-Lapsed Man and Other Stories* (1990) and *Blue Shifting* (1995), all from Pan Books, as well as over forty short stories to various markets such as *Zenith, Other Edens* and *Aboriginal SF*—including collaborative work with SF writers Stephen Baxter and Keith Brooke. His fiction has been translated into French, German, Italian, Czech and Rumanian. He has spent long periods travelling in India, Thailand and the Far East, and currently lives in Haworth, where he is player-manager of the Haworth Hackers soccer club. A novel for young adults, *Untouchable*, is forthcoming from Orion UK.

KIM NEWMAN

is the author of the novels *The Night Mayor, Bad Dreams, Jago, Anno Dracula, The Quorum* and *The Bloody Red Baron*. His short fiction is collected in *The Original Dr Shade* and *Famous Monsters*. He has also written and broadcast extensively about film, and his book-length non-fiction includes *Nightmare Movies* and *Wild West Movies*. His most recent works are *The British Film Institute Companion to Horror*, which he edited, and *Back in the USSA*, co-authored with Eugene Byrne. He is working on a complicated novel, *Life's Lottery*. "Great Western" takes place in an alternate-world version of the setting of *Jago*, which is roughly approximate to the stretch of Somerset where he grew up.

PETER F. HAMILTON

wrote his first short story in 1987, and is having a lot of trouble coming to terms with the fact he's been calling himself an author for ten years. His short fiction has appeared in *Fear*, *Interzone*, *In Dreams* and *New Worlds*. Several of these stories will be printed in his collection *A Second Chance at Eden* (Macmillan 1998). His first three novels, *Mindstar Rising*, *A Quantum Murder* and *The Nano Flower*, featuring the psychic detective Greg Mandel, are published by Tor in the USA; but his new trilogy is published by Warner Aspect, starting with *The Reality Dysfunction* in July 1997. This will be followed by *The Neutronium Alchemist* in 1998, and *The Naked God*, which according to the delivery date on his contract is coming out in 1999. He would like to point out that although his publisher's press release says his home is in Rutland Water, he does in fact live at the side of this fine reservoir.

GRAHAM JOYCE

quit an executive job and went to live on the Greek island of Lesbos, in a beach shack with a colony of scorpions (the setting for *House of Lost Dreams* [1993]), to concentrate on writing. He sold his first novel while still in Greece, and travelled in the Middle East on the proceeds. His other novels are *Dreamside* (1991), *Dark Sister* (1992), *Requiem* (1995) and *The Tooth Fairy* (1996), the last two published in the USA by Tor. *The Stormwatcher* is scheduled for summer 1997. Twice winner of the August Derleth Award (for *Dark Sister* and *Requiem*) presented by the British Fantasy Society, Joyce was also short-listed for the 1996 World Fantasy Award (for *Requiem*). His short stories have appeared in several anthologies, and his novels have been widely translated.

NOEL K. HANNAN

was born in 1967 and lives in Cheshire with his wife Helen, who is expecting their first child in July 1997. He is currently employed as a mainframe computer manager. "A Night on the Town" is his first professionally published story; however, his work has been extensively published both in the small press around the world and by comics companies, where his scripting credits include *Night of the Living Dead*, *Air Warriors* and *Streetmeat*. He runs a small press imprint, Bad to the Bone, and a website at http://ourworld.compuserve.com/homepages/BAD2BONE. Currently he is working on a novel, *Kingdoms of Clay*, a science fictionalisation of Antony and Cleopatra, featuring a race of alien cats. Any takers...?

BRIAN W. ALDISS

is President of APIUM, the Association for the Preservation and Integrity of an Unspoilt Mars. He is writing a utopia on this theme in collaboration with Roger Penrose (author of *The Emperor's New Mind*). His most recent books include *The Secret of This Book* (short stories), *The Detached Retina* (essays) and *At the Caligula Hotel* (poetry). His next book will be his autobiography, *The Twinkling of an Eye*, to be followed by the volumes of the *Squire Quartet*, revised and in a uniform edition. He has been writing for *New Worlds* since 1955.

ANDREW STEPHENSON

was born in Venezuela in 1946, because his parents happened to be there at the time. Other coincidences (too nasty to detail) led to an involvement in the community of SF fans in 1968 and his first short story being bought by John W. Campbell, Jr of *Analog* magazine in 1970. Other professional writing includes (indeed, virtually comprises) novels *Nightwatch* and *The Wall of Years* and a brace of short stories. SF is his favoured literary form. Having worked in telecommunications design, he can assure all who care to ask that imaginary gadgets are easier to debug; but an interest in real computer programming has stopped him getting cocky on that score. He hopes to commit that third novel Real Soon Now.

HOWARD WALDROP

was born in 1946, in Houston, Mississippi—but that didn't seem to help him any. He moved to Texas at age four, stuck around Dallas-Fort Worth area twenty years, and Austin for twenty-one. When he was forty-eight, he moved to Oso, Washington, to be two hundred yards from the trout. He has been a service-station flunkey, linotype operator, bad soldier, assistant editor of a literary magazine, term-paper-for-hire guy and guinea pig at an auditory research lab. All that time he was a writer, too. His most recent book is the collection *Going Home Again*, published in Australia. He's still finishing *I, John Mandeville* and *The Moon World*, both novels. And he's won some awards.

IAN WATSON's

last book was a technothriller about quantum computers, American cults and militias, *Hard Questions*. His next will be *Oracle*, in which a Roman centurion displaced from the time of Boadicea becomes involved with the IRA. He has recently been Guest of Honour at conventions in Belgium and Norway, and was also invited by the Swedish Academy of Sciences to a pow-wow in a hut beyond the Arctic Circle. He lives with his wife and two cats in a tiny village in the "empty quarter" of Northamptonshire, England (smaller than that of Arabia, and greener), within a stone's throw of the George Washington family shrine, Sulgrave Manor. A while ago he worked for nine months eyeball to eyeball with Stanley Kubrick on story development for the proposed SF movie *AI* (Artificial Intelligence). His daughter is a textile designer.

GARRY KILWORTH

was born in York city, England, of itinerant parents, and was raised partly in Aden. He has lived and worked in over twenty different countries throughout the world and has an honours degree in English from King's College, London. His latest books are *The Roof of Voyaging*, a Polynesian epic fantasy, and *A Midsummer's Nightmare*, a novel about the contemporary doings of Oberon, Titania and the other Dream fairies. "Attack of the Charlie Chaplins" also began as a dream. He lives on the hill overlooking Assundun, where Edmund Ironside was defeated by the Viking king Knut, way back in the Dark Ages when storytellers like him were hung on a gibbet for telling lies.

CHRISTINE MANBY

is the author of *Flatmates* (1997) and *Second Prize* (to be published in 1998). She sold her first short story at fourteen, but was a lowly audio-book production assistant when she chatted up David Garnett at a seedy publishing party in 1994 and he dared her to write a novel. She also worked in the music industry, but because there was not enough sex and drugs she became a freelance author. She has sold five books, three of which she has written as Stephanie Ash and is keeping from her parents because of their strong fantasy content. "For Life" is her first science fiction story. She used to run a dating agency, but now has only one client.

GRAHAM CHARNOCK

has been writing science fiction short stories for more than twenty-five years since his first appearance in *New Worlds* and has contributed to collections on both sides of the Atlantic. In the seventies he formed the cult British band Deep Fix with Michael Moorcock, later going on to front the proto-Goth group, Smackhead. More recently he has released three albums of "country swamp jazz" on his own Drop label and currently edits and publishes the bi-monthly fetish magazine devoted to body piercing and scarification, *Mutant Surgery*. He lives in North London with his partner, progeny and various animals.

MICHAEL MOORCOCK

works in most mediums and genres. Currently he's completed an ambitious CD-ROM interactive live-action story-game, is finishing the final volume of his Colonel Pyat sequence *The Vengeance of Rome*, beginning a London novel provisionally called *Sporting Club Square* and (with Simonson, Ridgeway and Reeve) producing a series of comics for DC called *Michael Moorcock's Multiverse*, which draws heavily on the Second Ether stories first published in *New Worlds*. Thanks to the wonders of modern electronics and applied Chaos Theory, he's comfortably settled in Texas, Spain, England and other parts of the multiverse.

WILLIAM GIBSON

has wanted to be a *New Worlds* writer since the very start of his career. His first professional sale ("The Gernsback Continuum," written in 1979) was in fact to *New Worlds*, but, via a preemptive strike by the late Terry Carr, and the kind understanding of Michael Moorcock, it appeared instead in Universe 11 (1981). He is delighted to at last appear in these pages. He thanks David Garnett for having extracted, before witnesses, in a pub in Charing Cross, the promise of a story.

DAVID GARNETT

has written some books, some stories, and edits *New Worlds*—and the second volume of this new series will be published by White Wolf.

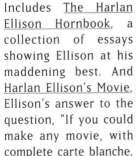